| Centre number | |
|---|---|
| Candidate number | |
| Surname and initials | |

CW01572886

 **Examining Group**

**General Certificate of Secondary Education**

# Mathematics
# Higher Tier
# Exam 1    Paper 1

## Time: two hours

### Instructions to candidates

Do **not** use a calculator.

Write your name, centre number and candidate number in the boxes at the top of this page.

Answer ALL questions in the spaces provided on the question paper.

Show all stages in any calculations and state the units.

Include diagrams in your answers where this may be helpful.

### Information for candidates

The number of marks available is given in brackets **[2]** at the end of each question or part question.

The marks allocated and the spaces provided for your answers are a good indication of the length of answer required.

| For Examiner's use only | |
|---|---|
| 1 | |
| 2 | |
| 3 | |
| 4 | |
| 5 | |
| 6 | |
| 7 | |
| 8 | |
| 9 | |
| 10 | |
| 11 | |
| 12 | |
| 13 | |
| 14 | |
| 15 | |
| 16 | |
| 17 | |
| **Total** | |

**EDUCATIONAL**

# Formulae Sheet: Higher Tier

**Volume of prism** = (area of cross section) × length

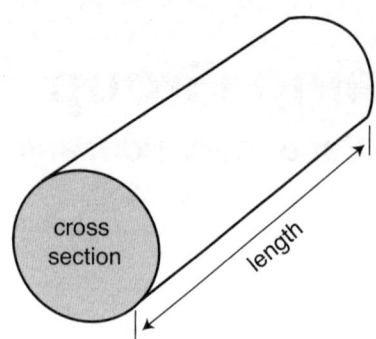

**In any triangle *ABC***

**Sine rule** $\dfrac{a}{\sin A} = \dfrac{b}{\sin B} = \dfrac{c}{\sin C}$

**Cosine rule** $a^2 = b^2 + c^2 - 2bc \cos A$

**Area of triangle** $= \dfrac{1}{2} ab \sin C$

**Volume of sphere** $= \dfrac{4}{3} \pi r^3$

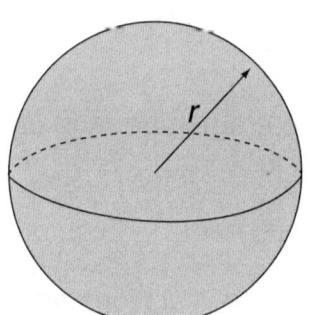

**Surface area of sphere** $= 4\pi r^2$

**Volume of cone** $= \dfrac{1}{3} \pi r^2 h$

**Curved area of cone** $= \pi r l$

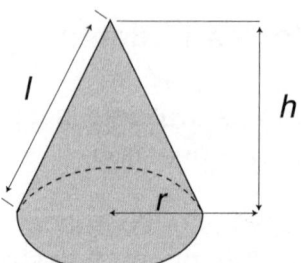

**The quadratic equation**

The solutions of $ax^2 + bx + c = 0$, where $a \neq 0$, are given by $x = \dfrac{-b \pm \sqrt{(b^2 - 4ac)}}{2a}$

*Letts*

**1** **(a)** $136 \times 57 = 7752$
Find:

    **(i)** $0.0136 \times 570$ ...............................................................................

    **(ii)** $0.7752 \div 13.6$ ............................................................................... **[2]**

**(b)** Write 210 as a product of its prime factors.

.........................................................................................................

......................................................................................................... **[2]**

**(Total 4 marks)**

**2** Solve these equations.

**(a)** $7x - 4(x - 3) = 27$

.........................................................................................................

......................................................................................................... **[3]**

**(b)** $\dfrac{x + 3}{2} - \dfrac{x - 4}{5} = 5$

.........................................................................................................

......................................................................................................... **[4]**

**(Total 7 marks)**

**3** Simplify the expression:

$$\frac{(36 \times 10^{-9}) \times (5 \times 10^{6})}{48 \times 10^{2}}$$

Give your answer in standard form.

.........................................................................................................

......................................................................................................... **[3]**

**(Total 3 marks)**

*Letts*

**4**   The diagram shows two straight roads crossing at A.
A cellular telephone company is going to erect a mast.
It must be:
- nearer B than A;
- nearer AC than AB;
- less than 6 km from A.

Draw these boundaries accurately and shade the region where the mast can be.

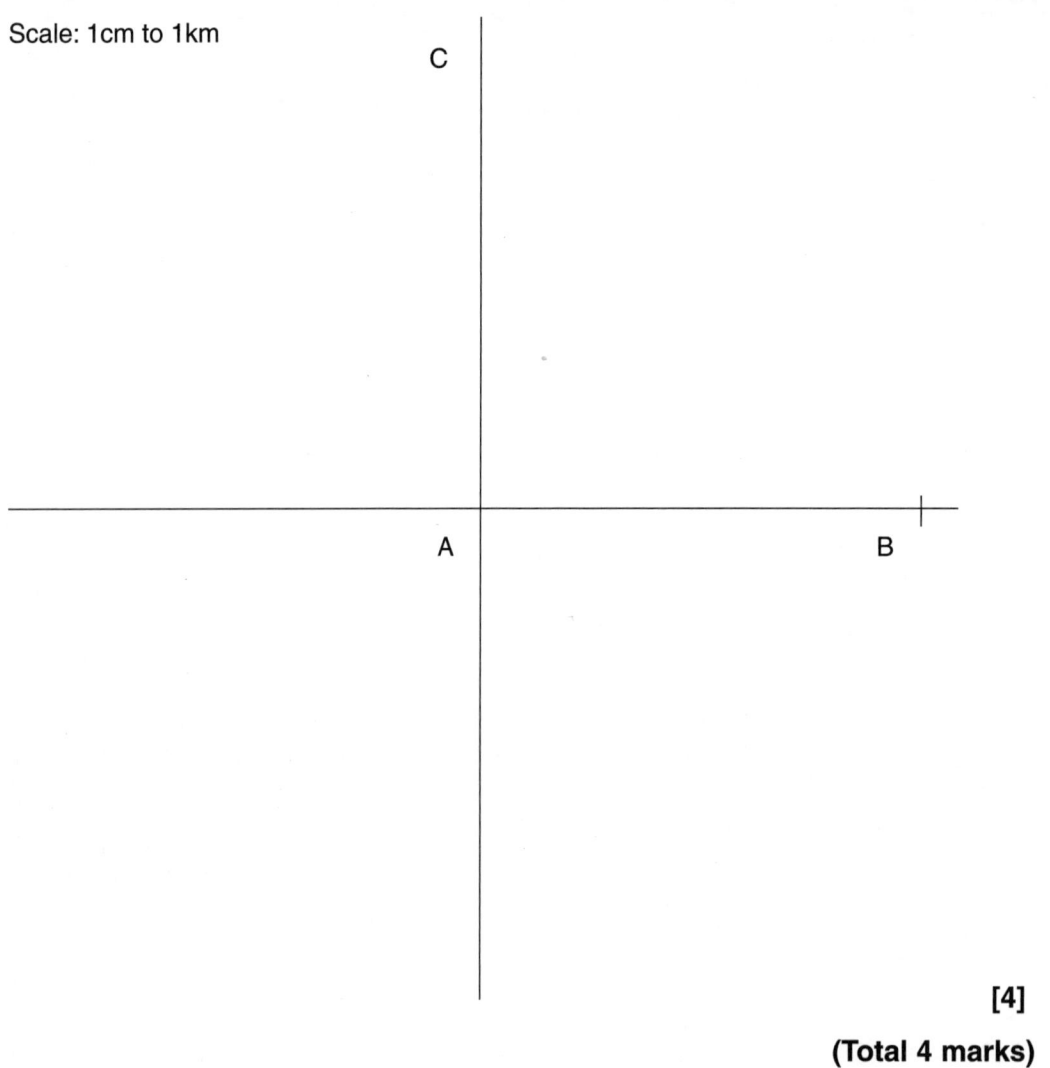

Scale: 1cm to 1km

**[4]**

**(Total 4 marks)**

**5** **(a)** On the axes below, draw the graph of:

$$y = x^3 - 3x + 3$$

from $x = -2$ to $x = 2$.

...................................................................................................

...................................................................................................

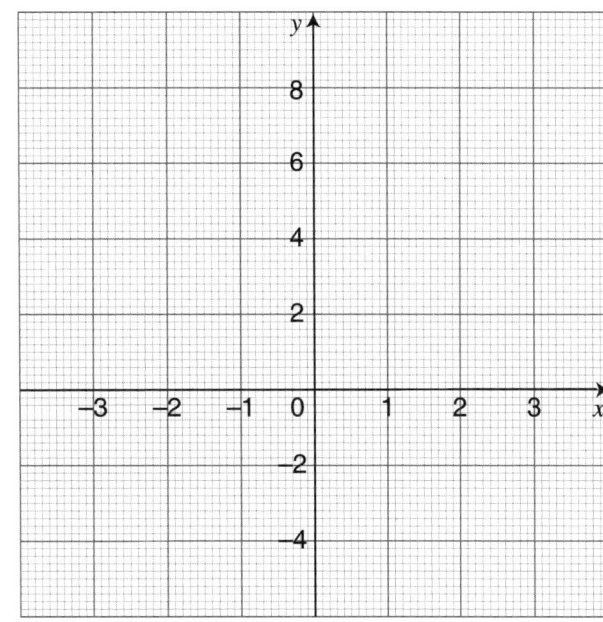

[4]

**(b)** **(i)** On the same axes, draw the graph of $y = 3$.
Write down the $x$-coordinates of the points where the two graphs meet.

...................................................................................................

...................................................................................................

**(ii)** Find the equation for which these are the solutions.

...................................................................................................

................................................................................... [3]

**(Total 7 marks)**

**6**   TB is a tangent to the circle centre O.
Prove that angle AOB is $2x°$.

..........................................................................................................

..........................................................................................................

..........................................................................................................

.......................................................................................... **[4]**

**(Total 4 marks)**

**7** The formula:

$$d = An^2 + Bn$$

gives the number of diagonals, $d$, that can be drawn in a polygon with $n$ sides.

**(a)** By considering the number of diagonals in a triangle and a quadrilateral, show that:

$$3A + B = 0$$

$$8A + 2B = 1$$

........................................................................................................................

........................................................................................................................ **[2]**

**(b)** Use algebra to find the values of $A$ and $B$.

........................................................................................................................

........................................................................................................................

........................................................................................................................ **[3]**

**(c)** Hence find the number of diagonals in a polygon with 15 sides.

........................................................................................................................ **[1]**

**(Total 6 marks)**

**[turn over**

**8**

Not to scale

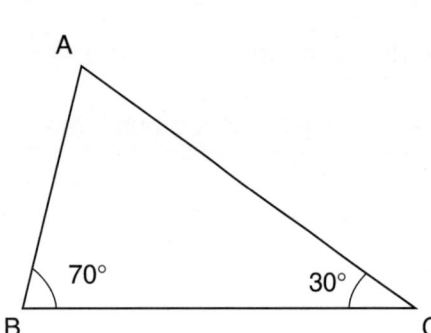

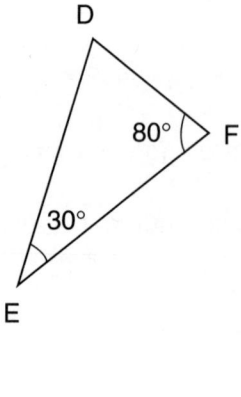

**(a)** Explain why these triangles are similar.

..............................................................................................................

.............................................................................................................. **[1]**

**(b)** AC = 12 cm, BC = 9 cm, DF = 2 cm, EF = 4 cm.

Find the lengths of AB and DE.

..............................................................................................................

..............................................................................................................

..............................................................................................................

.............................................................................................................. **[4]**

**(Total 5 marks)**

**9** A council spends money on Education, Social Services and Other Services in the ratio 7 : 2 : 3.
£420 million was spent on Education, how much was spent on Other Services?

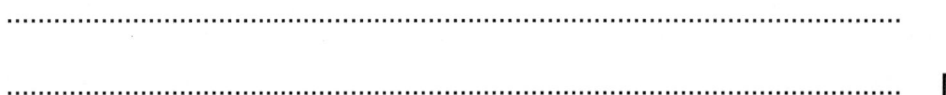

.............................................................................................................. **[3]**

**(Total 3 marks)**

**10** The population of a village is 3000.
The histogram shows the distribution of the ages of the population.

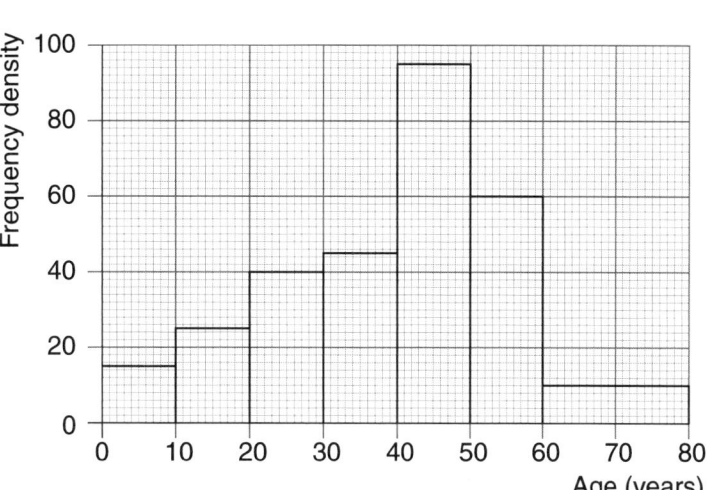

(a) Find the values of *r* and *s* in the frequency table below.

| Age (*a* years) | 0<a≤10 | 10<a≤20 | 20<a≤30 | 30<a≤40 | 40<a≤50 | 50<a≤60 | 60<a≤80 |
|---|---|---|---|---|---|---|---|
| Frequency | 150 | 250 | 400 | 450 | 950 | *r* | *s* |

[2]

(b) Complete the cumulative frequency table below.

| Age in years | ≤10 | ≤20 | ≤30 | ≤40 | ≤50 | ≤60 | ≤80 |
|---|---|---|---|---|---|---|---|
| Cumulative frequency | 150 | 400 | | | | | 3000 |

[1]

(c) On the grid below, draw a cumulative frequency graph to illustrate this information.

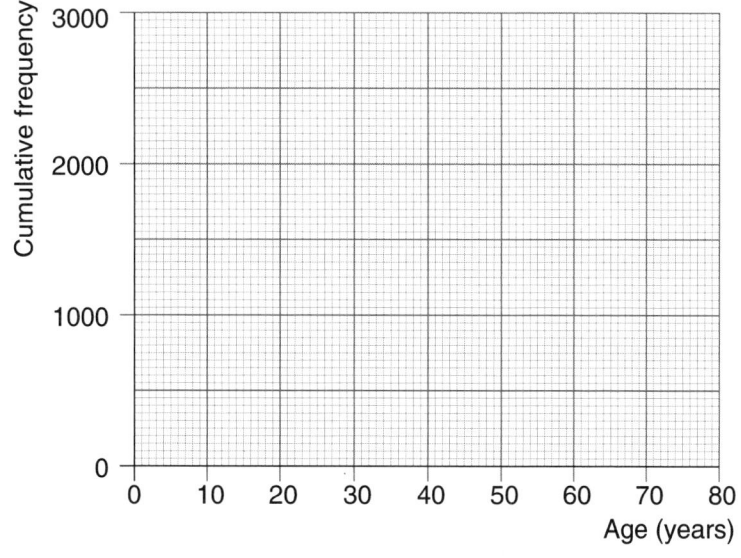

[2]

**(d)** Use your graph to estimate:

    **(i)** the number of people who are more than 57 years old,

    **(ii)** the median age of the population.

    **(i)** .............................................................................................

    **(ii)** .......................................................................................... **[3]**

                                                **(Total 8 marks)**

**11** In any triangle the sum of the lengths of two of the sides must be greater than the length of the third side.
For the triangle below find the range of values of $x$ for this statement to be always true.

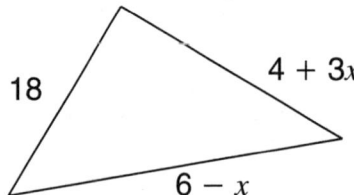

.................................................................................................... **[5]**

                                                **(Total 5 marks)**

**12**

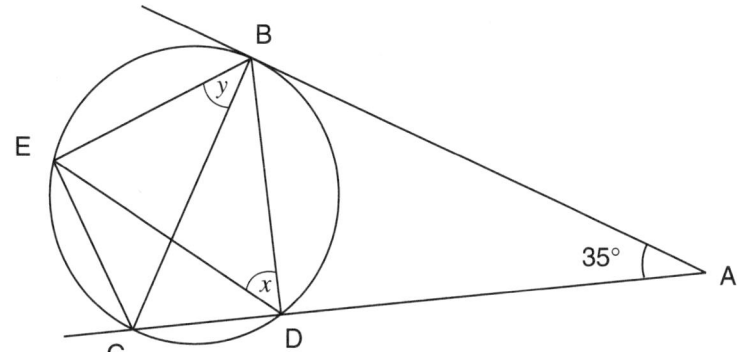

AB is a tangent and BC a diameter to the circle.
Angle BAC = 35°.

**(a)** Find two other angles equal to 35°.

..................................................................................................................... **[2]**

**(b)** Find a relationship between $x$ and $y$.
Give a reason for each step of your argument.

....................................................................................................................

....................................................................................................................

..................................................................................................................... **[3]**

**(c)** Find three similar triangles.

..................................................................................................................... **[2]**

**(Total 7 marks)**

**13** Find the points of intersection of the straight line:

$$y = x - 7$$

with the circle:

$$x^2 + y^2 = 25$$

..............................................................................................................

..............................................................................................................

..............................................................................................................

..............................................................................................................

..............................................................................................................

..............................................................................................................

..............................................................................................................

..............................................................................................................

.............................................................................................................. **[6]**

**(Total 6 marks)**

**14 (a)** Solve this equation:

$$2x^2 - 5x - 3 = 0$$

...................................................................................................

...................................................................................................

...................................................................................................

................................................................................................... **[3]**

**(b)** Simplify and factorise:

$$y(2 - y)(3 - y) + y^2(y + 5).$$

...................................................................................................

...................................................................................................

...................................................................................................

................................................................................................... **[3]**

**(c)** Rearrange the formula:

$$h = \frac{3u^2}{8g}$$

to make $u$ the subject.

...................................................................................................

...................................................................................................

...................................................................................................

................................................................................................... **[3]**

**(Total 9 marks)**

    **[turn over**

**15** There are five identical tins, without labels, on a shelf.
Three tins contain baked beans, the other two contain peas.

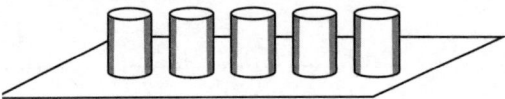

John takes two tins.

**(a)** What is the probability that they both contain baked beans?
Leave your answer as a fraction.

.............................................................................................................

.............................................................................................................

.............................................................................................................

............................................................................................. **[3]**

John does not open the tins and does not put them back.
He takes one of the remaining tins.

**(b)** What is the probability that it contains peas?
Leave your answer as a fraction.

.............................................................................................................

.............................................................................................................

.............................................................................................................

............................................................................................. **[5]**

**(Total 8 marks)**

**16**

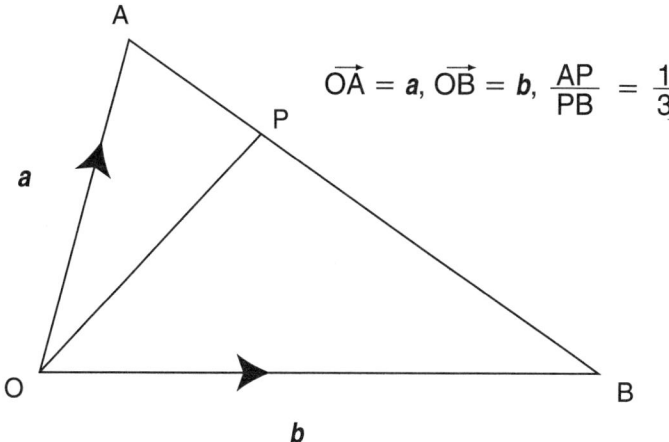

$\overrightarrow{OA} = a$, $\overrightarrow{OB} = b$, $\dfrac{AP}{PB} = \dfrac{1}{3}$

Leave blank

**(a)** Find, in terms of **a** and **b**, **(i)** $\overrightarrow{AB}$ **(ii)** $\overrightarrow{PB}$ **(iii)** $\overrightarrow{OP}$

..................................................................................................................

..................................................................................................................

.................................................................................................... **[4]**

**(b)** Q is the point such that OAQP is a parallelogram.
Find $\overrightarrow{BQ}$ in terms of **a** and **b**.

..................................................................................................................

..................................................................................................................

..................................................................................................................

.................................................................................................... **[5]**

**(Total 9 marks)**

**17** A whole number has digits $a$, $b$ and $c$ (in that order).

**(a)** Write an algebraic expression to represent this number.

............................................................................................................................... **[2]**

**(b)** Prove that the difference between a 3-digit number and another number, which has the same digits but reversed, is divisible by 9.

............................................................................................................................

............................................................................................................................

............................................................................................................................

............................................................................................................................

............................................................................................................................... **[3]**

**(Total 5 marks)**

 **Examining Group**

**General Certificate of Secondary Education**

# Mathematics
# Higher Tier
# Exam 1    Paper 2

## Time: two hours

### Instructions to candidates

You **may** use a calculator.

Write your name, centre number and candidate number in the boxes at the top of this page.

Answer ALL questions in the spaces provided on the question paper.

Show all stages in any calculations and state the units.

Include diagrams in your answers where this may be helpful.

### Information for candidates

The number of marks available is given in brackets **[2]** at the end of each question or part question.

The marks allocated and the spaces provided for your answers are a good indication of the length of answer required.

**EDUCATIONAL**

# Formulae Sheet: Higher Tier

**Volume of prism** = (area of cross section) × length

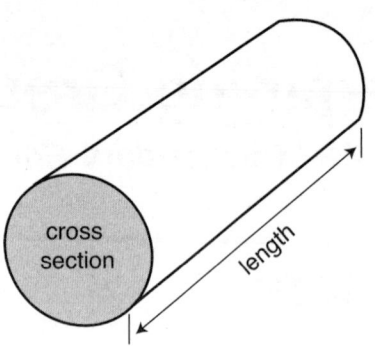

**In any triangle *ABC***

**Sine rule** $\dfrac{a}{\sin A} = \dfrac{b}{\sin B} = \dfrac{c}{\sin C}$

**Cosine rule** $a^2 = b^2 + c^2 - 2bc \cos A$

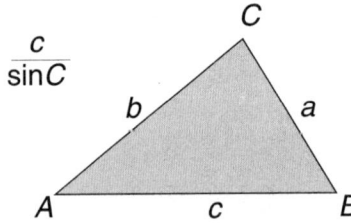

**Area of triangle** $= \dfrac{1}{2} ab \sin C$

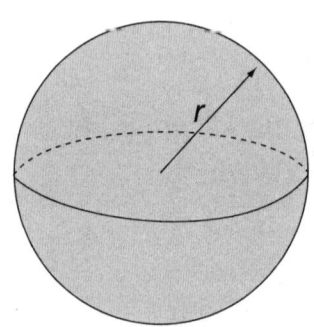

**Volume of sphere** $= \dfrac{4}{3} \pi r^3$

**Surface area of sphere** $= 4\pi r^2$

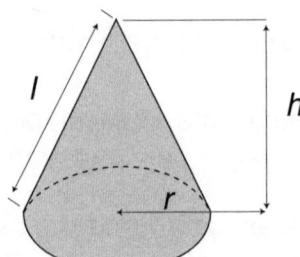

**Volume of cone** $= \dfrac{1}{3} \pi r^2 h$

**Curved area of cone** $= \pi r l$

**The quadratic equation**

The solutions of $ax^2 + bx + c = 0$, where a ≠ 0, are given by $x = \dfrac{-b \pm \sqrt{(b^2 - 4ac)}}{2a}$

1    The table shows data kept by a public library.

| Year | 1969 | 1999 |
|---|---|---|
| Number of books in stock | 16 539 | 210 420 |
| Number of readers | 4861 | 25 817 |
| Number of borrowings | 127 592 | 1 262 837 |

Comment, showing your calculations.

.........................................................................................................

.........................................................................................................

.........................................................................................................

.................................................................................................... **[3]**

**(Total 3 marks)**

2    The box plots show information about the heights in metres of a group of boys and girls.

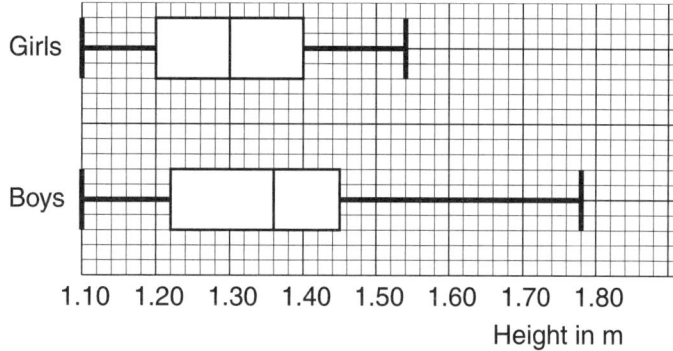

Height in m

(a) Complete the table.

|  | Median | Interquartile range | Tallest height |
|---|---|---|---|
| Girls |  |  |  |
| Boys |  |  |  |

**[4]**

(b) Make two comparisons.

.........................................................................................................

.................................................................................................... **[2]**

**(Total 6 marks)**

**3** This is the graph of $y = x^3 - 2x^2 - 5x + 5$.

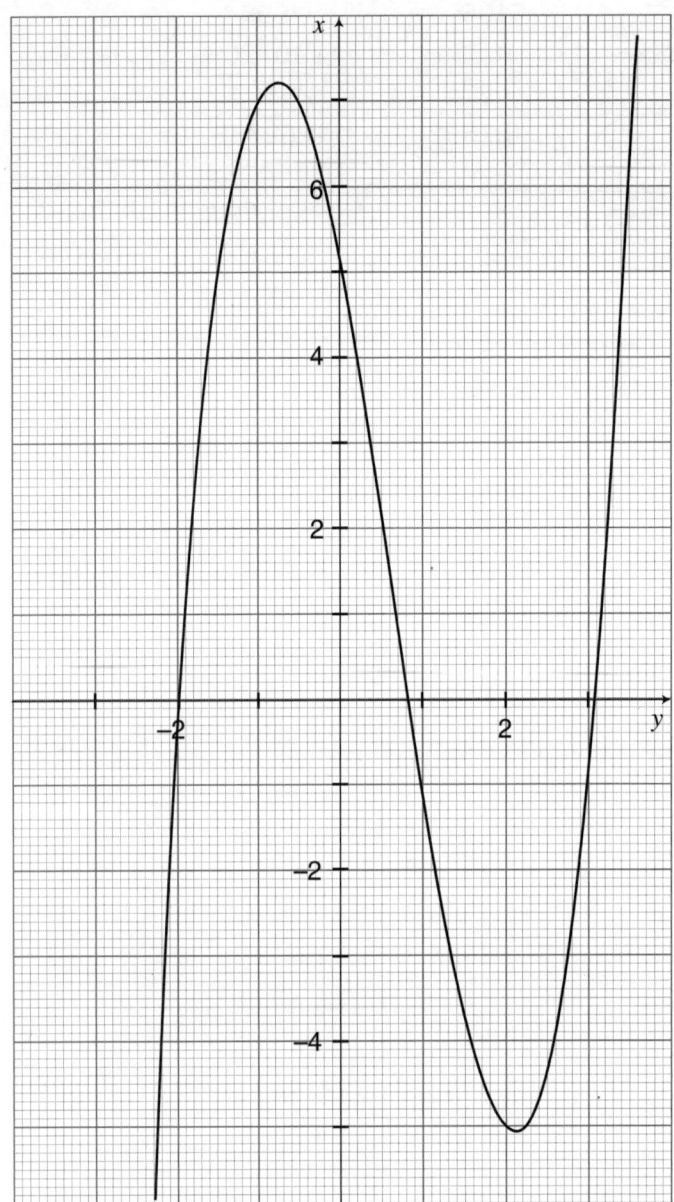

**(a)** Use the graph to solve the equation.

$$x^3 - 2x^2 - 5x + 5 = 0$$

............................................................................................................. **[2]**

**(b)** Use trial and improvement to find the largest root correct to 2 decimal places.

.........................................................................................................

.........................................................................................................

.........................................................................................................

............................................................................................................. **[4]**

**(Total 6 marks)**

© Letts Educational 2003                    4

*Letts*

**4**   Calculate the area of this trapezium.  The lengths are in centimetres.

.............................................................................................................

.............................................................................................................

.............................................................................................................

.......................................................................................................... **[5]**

**(Total 5 marks)**

**5**   Assume that you are 16 years old today.
Assume that it takes you 2 minutes to do this question.
What fraction of your life have you spent doing this question?
Give you answer in standard form to a suitable degree of accuracy.

.............................................................................................................

.............................................................................................................

.............................................................................................................

.......................................................................................................... **[4]**

**(Total 4 marks)**

**6** AB is a straight line which passes through the points (2, 5) and (6, 13).

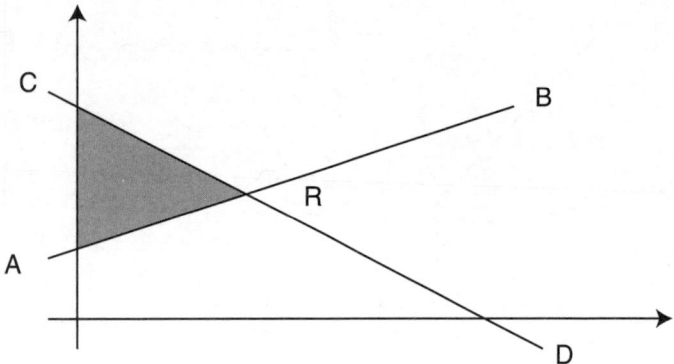

(a) What is the equation of the line AB?

..................................................................................................

..................................................................................................

..................................................................................................

.................................................................................................. **[4]**

(b) Another line, CD, meets AB at R.
The equation of CD is $y = 4 - x$.
Calculate the area between the two lines and the $y$ axis (shaded on the diagram).

..................................................................................................

..................................................................................................

..................................................................................................

.................................................................................................. **[4]**

**(Total 8 marks)**

**7** Anne and Brian are about to take their driving tests.
The probability that Anne will pass is 0.65.
The probability that Brian will pass is 0.4.
These two events are independent.

**(a)** Draw a tree diagram to show the outcomes.

[3]

**(b)** Find the probabilities that:

**(i)** they both pass, **(ii)** only one of them passes.

.........................................................................................................

.........................................................................................................

.........................................................................................................

......................................................................................................... [4]

**(Total 7 marks)**

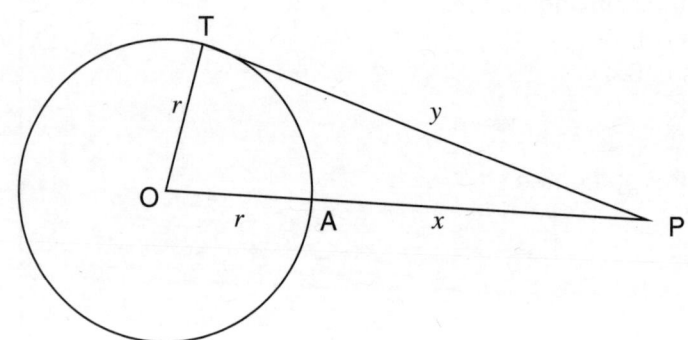

TP is a tangent to the circle centre O and OAP is a straight line.

**(a)** Find a formula connecting $y$ with $x$ and $r$ in its simplest form.

...................................................................................................

................................................................................................... **[3]**

**(b)** If $r = 6340$ and $y = 150$, find $x$.

...................................................................................................

...................................................................................................

................................................................................................... **[4]**

**(c) (i)** If $x$ is very small compared with $r$ and $y$, explain why $y \approx \sqrt{2rx}$ .

...................................................................................................

................................................................................................... **[1]**

**(ii)** The radius of the Earth is approximately 6340 km.
Use the formula in (i) to find how far it is to the horizon from the top of a mountain 1200 m high.

...................................................................................................

...................................................................................................

................................................................................................... **[2]**

**(Total 10 marks)**

**9** Mr Brown invested £12000 at 6.5% per annum compound interest.

    **(a)** How much will he have after 5 years?

        .................................................................................................

        .................................................................................... **[2]**

    **(b)** After that he invests a further £3000.

        How many years will it be before the total becomes greater than £24000?

        .................................................................................................

        .................................................................................................

        .................................................................................................

        .................................................................................... **[4]**

                                                      **(Total 6 marks)**

10 (a) Melted wax is poured into a cylindrical mould with capacity 125 cm³.
The height of the cylinder is 12.0 cm.
Calculate its radius.

..............................................................................................

.......................................................................................... [3]

(b) A mould in the shape of a cone also has capacity 125 cm³.
Its height is equal to the diameter of the base.
Calculate its height.

..............................................................................................

.......................................................................................... [3]

(c) Another 125 cm³ of wax is poured into conical moulds.
These moulds are mathematically similar to the mould in (b) but
half the height.
How many will be filled?

..............................................................................................

.......................................................................................... [2]

**(Total 8 marks)**

11

$$T = \frac{100(P - A)}{AR}$$

Rearrange this formula to make A the subject.

..............................................................................................

..............................................................................................

..............................................................................................

.......................................................................................... [4]

**(Total 4 marks)**

**12** Wooden models are made in the form of truncated square pyramids as shown.

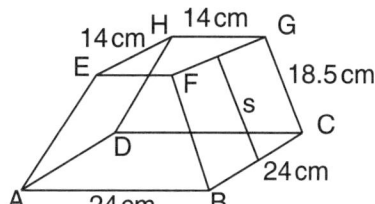

The base, ABCD has sides 24.0 cm long, the slant edge, GC, is 18.5 cm and the top, EFGH, has sides 14 cm long.

**(a)** Calculate

**(i)** the vertical height,

........................................................................................................

........................................................................................................ **[4]**

**(ii)** the slant height, *s*,

........................................................................................................

........................................................................................................ **[2]**

**(iii)** the angle between the base and the slant height, *s*.

........................................................................................................

........................................................................................................ **[2]**

**(iv)** Find the total surface area of the model.

........................................................................................................ **[2]**

**(Total 10 marks)**

**13** Solve the equation:

$$\frac{3}{x-1} - \frac{1}{2x+1} = 1$$

....................................................................................................

....................................................................................................

....................................................................................................

.................................................................................................... **[6]**

**(Total 6 marks)**

**14** The graph shows ice cream sales for 2001 and 2002.

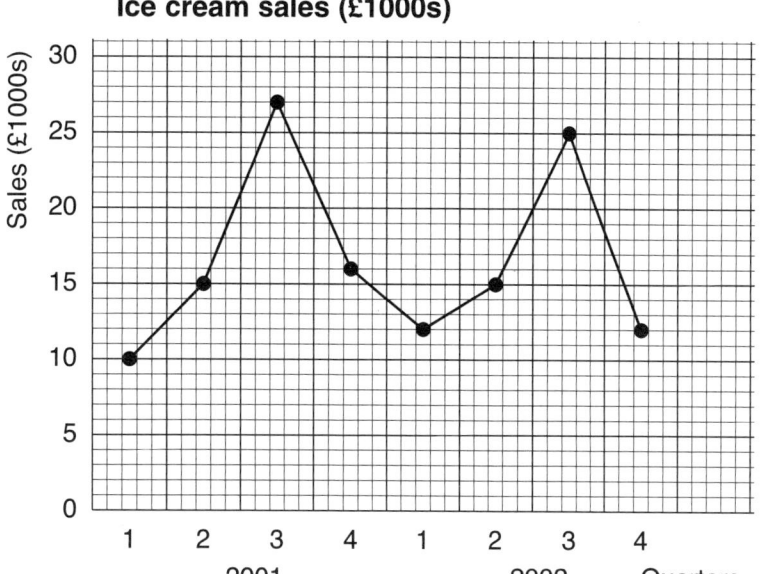

**Ice cream sales (£1000s)**

A four-quarter moving average is to be found.

**(a)** Why is a four-quarter average used?

..................................................................................................... [1]

**(b)** Find the moving averages and plot them on the graph.

.....................................................................................................

..................................................................................................... [3]

**(c)** Find the sales in the first quarter of 2003 if the next moving average is 15.5.

.....................................................................................................

..................................................................................................... [1]

**(Total 5 marks)**

**[turn over**

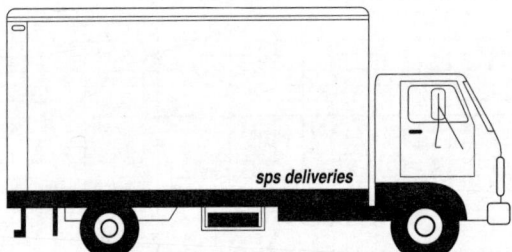

A truck can safely carry 15 tonnes.
It is loaded with cubical blocks measuring $300 \times 100 \times 100$ mm.
The density of the blocks is $1500$ kg/m$^3$.
All the measurements are correct to 2 significant figures.

Find the maximum number of blocks the truck can safely carry.

...................................................................................................................

...................................................................................................................

...................................................................................................................

................................................................................................................... [5]

**(Total 5 marks)**

**16** Simplify:

**(a)** $\dfrac{x^2 - 9}{3x^2 - 7x - 6}$

.......................................................................................................

.......................................................................................................

....................................................................................................... **[4]**

**(b)** $\dfrac{3}{x + 4} + \dfrac{4}{x}$

.......................................................................................................

.......................................................................................................

....................................................................................................... **[3]**

**(Total 7 marks)**

**[turn over**

*Letts*

**BLANK PAGE**

# Examining Group

## General Certificate of Secondary Education

# Mathematics
# Higher Tier
# Exam 2    Paper 1

## Time: two hours

### Instructions to candidates

Do **not** use a calculator.

Write your name, centre number and candidate number in the boxes at the top of this page.

Answer ALL questions in the spaces provided on the question paper.

Show all stages in any calculations and state the units.

Include diagrams in your answers where this may be helpful.

### Information for candidates

The number of marks available is given in brackets **[2]** at the end of each question or part question.

The marks allocated and the spaces provided for your answers are a good indication of the length of answer required.

EDUCATIONAL

## Useful Formulae

**Volume of prism** = (area of cross section) × length

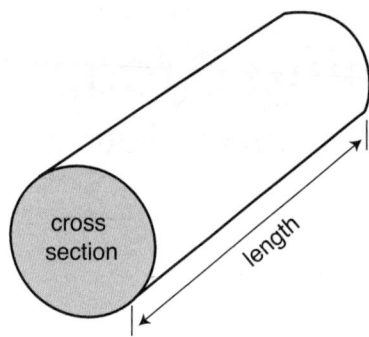

**In any triangle *ABC***

**Sine rule** $\dfrac{a}{\sin A} = \dfrac{b}{\sin B} = \dfrac{c}{\sin C}$

**Cosine rule** $a^2 = b^2 + c^2 - 2bc\cos A$

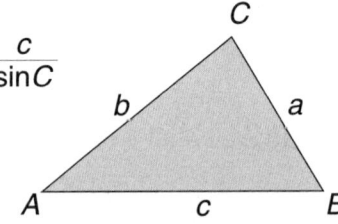

**Area of triangle** $= \dfrac{1}{2}ab\sin C$

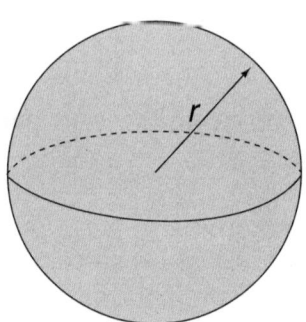

**Volume of sphere** $= \dfrac{4}{3}\pi r^3$

**Surface area of sphere** $= 4\pi r^2$

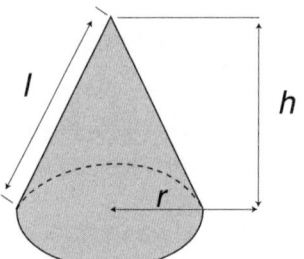

**Volume of cone** $= \dfrac{1}{3}\pi r^2 h$

**Curved area of cone** $= \pi r l$

**The quadratic equation**

The solutions of $ax^2 + bx + c = 0$, where a ≠ 0, are given by $x = \dfrac{-b \pm \sqrt{(b^2 - 4ac)}}{2a}$

*Letts*

**1** Work out an estimate for the value of:

$$\frac{62.5 \times 5.07 - 9.89 \times 3.06}{97.8^2}$$

Give your answer as a fraction in its simplest form.

.................................................................................................

.................................................................................................

.................................................................................................

.................................................................................................

................................................................................. **[3]**

**(Total 3 marks)**

**2** Here are the first five terms of an arithmetic sequence.

7, 10, 13, 16, 19

Find an expression in terms of $n$, for the $n$th term of the sequence.

.................................................................................................

................................................................................. **[2]**

**(Total 2 marks)**

**3** Using the information that:

$28 \times 321 = 8988$

Write down the value of:

**(i)** $2.8 \times 321$ ...........................................................................

................................................................................. **[1]**

**(ii)** $28 \times 0.321$ .........................................................................

................................................................................. **[1]**

**(iii)** $89.88 \div 2.8$ .........................................................................

................................................................................. **[1]**

**(Total 3 marks)**

**4**   On the grid triangle B is the image of triangle A after reflection.

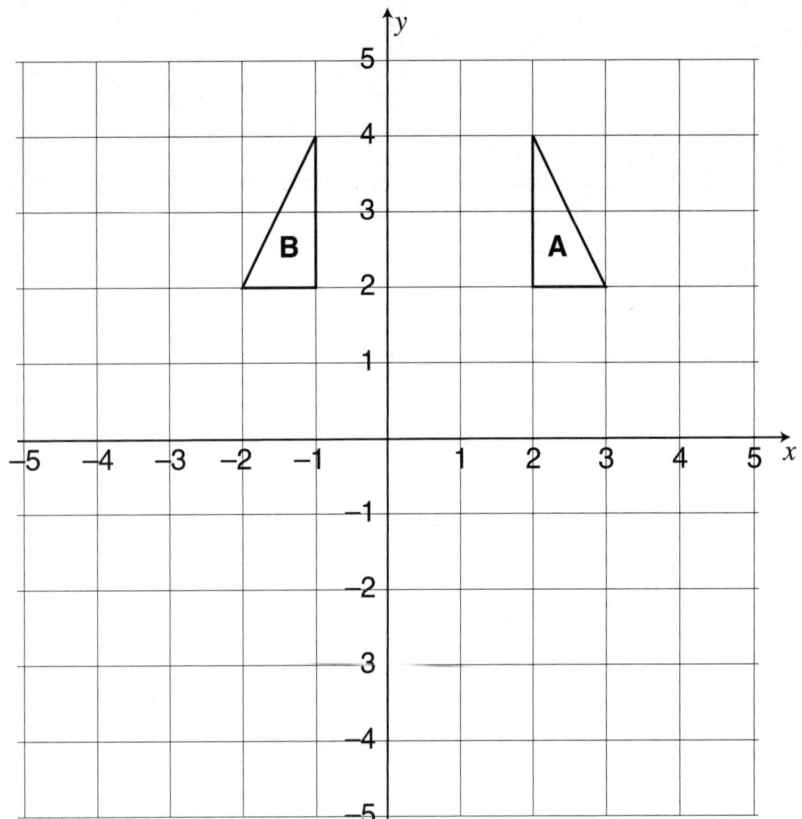

(a)  Write down the equation of the line of reflection.

......................................................................................................................... [1]

(b)  Rotate triangle A through 90° clockwise about (2,1)

[2]

(c)  Translate triangle B by the vector $\begin{pmatrix} -2 \\ -4 \end{pmatrix}$

[2]
**(Total 5 marks)**

**5**  Solve $6a - 3 = 3(a - 5)$

.........................................................................................................................

.........................................................................................................................

$a$  ................................ **[2]**
**(Total 2 marks)**


**6**  The table shows some expressions.

The letters $a$, $b$ and $c$ represent lengths.

Place a tick in the appropriate column for each expression to show whether the expression can be used to represent a length, an area, a volume or none of these.

| Expression | Length | Area | Volume | None of these |
|---|---|---|---|---|
| $a^2 + bc$ | | | | |
| $\dfrac{abc}{\pi b^2}$ | | | | |
| $3a^2\sqrt{b^2 + c^2}$ | | | | |

**[3]**
**(Total 3 marks)**

**7**  The times in minutes taken by 11 people to wait to be served at a supermarket checkout are listed in order:

1, 1, 2, 3, 3, 3, 4, 5, 5, 6, 8

**(a)** Find:

**(i)** the lower quartile

.......................................................................................................................

........................................................................................ minutes

**(ii)** the interquartile range

.......................................................................................................................

........................................................................................ minutes    **[3]**

**(b)** Draw a box plot for this data on the grid below.

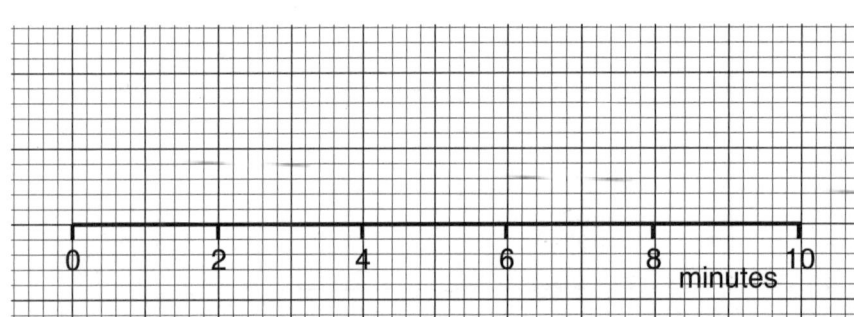

**[3]**
**(Total 6 marks)**

**8** $a = 6 \times 10^7$

$b = 3 \times 10^4$

**(a)** Find the value of $a \times b$

Give your answer in standard form.

........................................................................................................................

........................................................................................................................ **[2]**

**(b)** Find the value of $\dfrac{a^2}{b}$

Give your answer in standard form.

........................................................................................................................

........................................................................................................................

........................................................................................................................ **[3]**

**(Total 5 marks)**

**9** **(a)** Simplify:

   **(i)** $\dfrac{n^6}{n^2}$ ...................................................................................... [1]

   **(ii)** $\dfrac{2h^4 \times 3h^7}{9h^{12}}$ ............................................................................ [1]

**(b)** Expand and simplify:

   **(i)** $(5x - 3)(2x + 7)$

   .................................................................................................. [2]

   **(ii)** $(5x - 4)^2$

   .................................................................................................. [3]

**(c)** Solve the equation:

   $x^2 + 4x - 12 = 0$

   .................................................................................................. [3]
   **(Total 10 marks)**

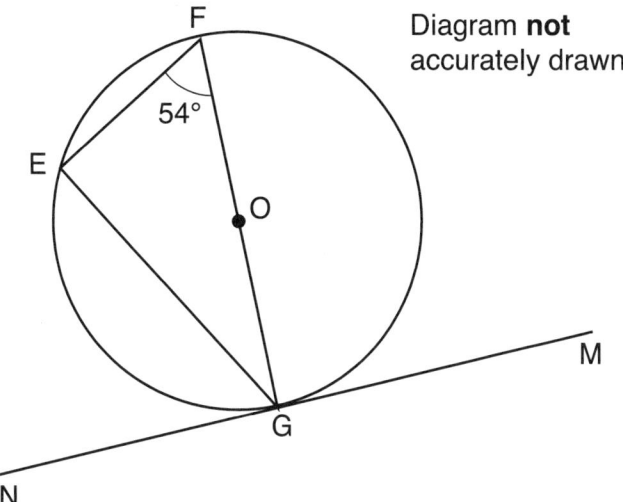

Diagram **not** accurately drawn

In the diagram E, F and G are points on the circle, centre O.

Angle EFG is 54°.

MN is a tangent to the circle at point G.

    **(i)** Calculate the size of angle FGE.
        Give reasons for your answer.

.................................................................................................................

.................................................................................................................

.................................................................................................. °   **[2]**

    **(ii)** Calculate the size of angle EGN.
        Give reasons for your answer.

.................................................................................................................

.................................................................................................................

.................................................................................................. °   **[2]**

                                                  **(Total 4 marks)**

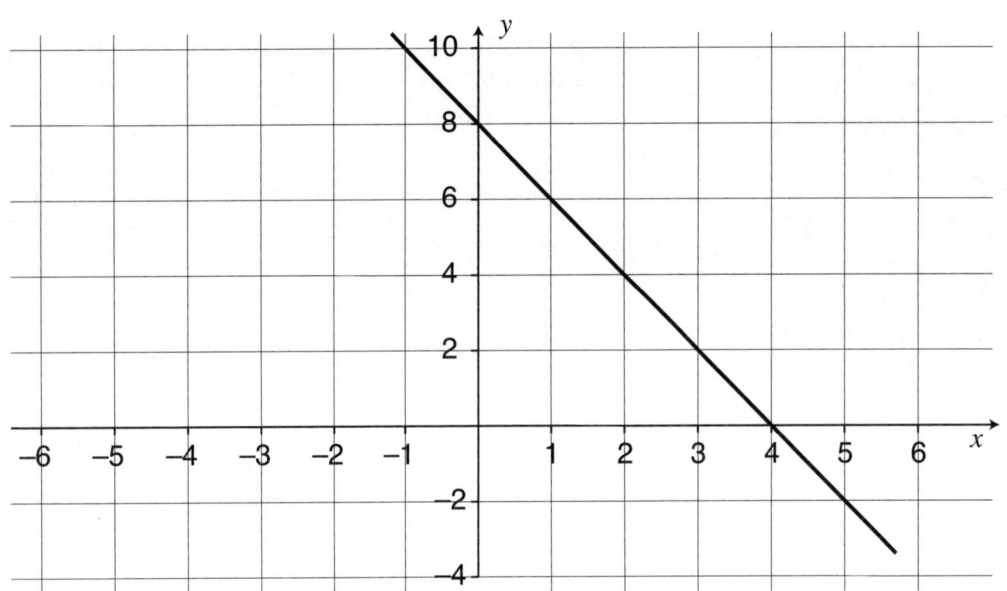

A straight line has been drawn on the grid.

**(a)** Write down the equation of the straight line.

...........................................................................................................

...........................................................................................................

$$y = \quad .....................$$  **[2]**

**(b)** Write down the gradient of the line    $x + 3y = 6$.

...........................................................................................................

........................................................................................................... **[2]**

**(c)** Write down the equation of the line which is parallel to the line with the equation $4x + 2y = 8$ and passes through the point with coordinates $(0, -1)$.

...........................................................................................................

...........................................................................................................

........................................................................................................... **[2]**

**(d)** Write down the equation of a line which is perpendicular to the line $y = 3x$.

...........................................................................................................

........................................................................................................... **[2]**

**(Total 8 marks)**

**12** Work out:

   **(i)** $5^0$ ................................................................................ [1]

   **(ii)** $8^{-2}$ ............................................................................ [1]

   **(iii)** $27^{\frac{2}{3}}$ ............................................................................ [1]

   **(iv)** $\frac{1}{25}^{-\frac{1}{2}}$ ............................................................................ [2]

   **(Total 5 marks)**

**13** Rearrange the formula $y = \dfrac{a(x + b)}{x - c}$

   to make $x$ the subject.

   ..............................................................................................

   ..............................................................................................

   ..............................................................................................

   ..............................................................................................

   ............................................................................................... [4]

   **(Total 4 marks)**

*Letts*

**14** A bag contains 4 red beads, 2 black beads and 3 green beads.
Rosie takes a bead at random from the bag, records its colour and replaces it.
She does this two more times.

Work out the probability that of the three beads Rosie takes out at least two are the same colour.

.......................................... **[6]**
**(Total 6 marks)**

**15** Work out:

$$\frac{(7 + \sqrt{5})\ (7 - \sqrt{5})}{\sqrt{80}}$$

Leave your answer in the form $\frac{a\sqrt{b}}{c}$ .

......................................................................................................

......................................................................................................

......................................................................................................

.................................................................................... **[4]**
**(Total 4 marks)**

**16** $a$ is directly proportional to the square of $b$.

When $a = 12, b = 2$.

**(a)** Find an expression for $a$ in terms of $b$.

...................................................................................................

...................................................................................................

$a =$ ................................. **[3]**

**(b)** Calculate $a$ when $b = 3$.

...................................................................................................

...................................................................................................

...................................................................................................  **[1]**

**(c)** Calculate $b$ when $a = 192$.

...................................................................................................

...................................................................................................

...................................................................................................  **[2]**

(Total 6 marks)

**17** The table and histogram gives information about how long in minutes, some students took to complete a puzzle.

| Time (t) in minutes | Frequency |
|---|---|
| $0 < t \leqslant 5$ | ……… |
| $5 < t \leqslant 15$ | 52 |
| $15 < t \leqslant 30$ | 48 |
| $30 < t \leqslant 50$ | 88 |
| $50 < t \leqslant 60$ | ……… |

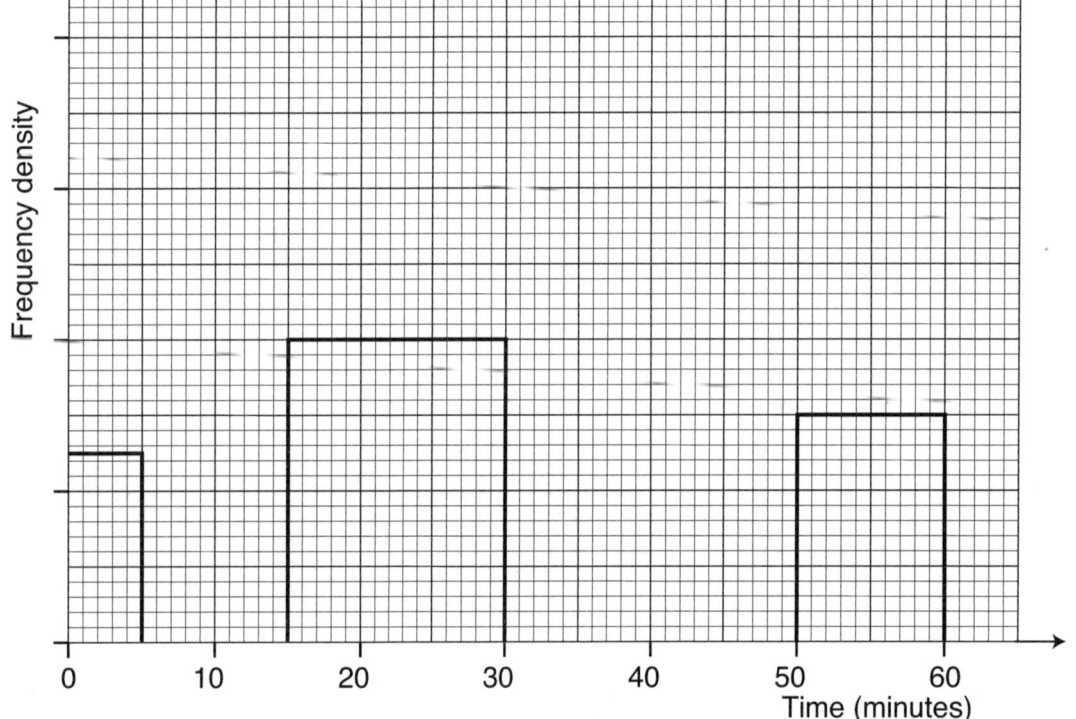

**(a)** Use the information in the histogram to complete the table. **[2]**

**(b)** Use the table to complete the histogram. **[2]**

**(Total 4 marks)**

**18** The diagram shows a toy.
The toy is made up of a cone and a hemisphere.
Work out the volume of the toy.
Leave your answer in terms of π.

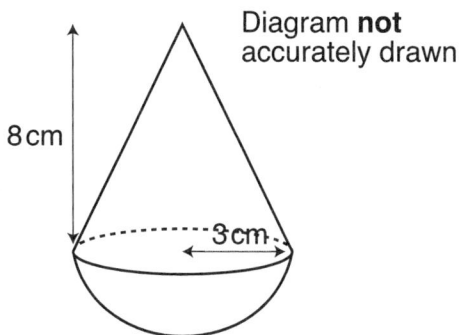

Diagram **not**
accurately drawn

8 cm

3 cm

.................... cm³   **[7]**
**(Total 7 marks)**

**19** Simplify fully:

$$\frac{2n^2 + n - 6}{4n^2 - 9} \times \frac{4n + 6}{n^2 + 3n + 2}$$

**[6]**
**(Total 6 marks)**

**20** In this diagram $\overrightarrow{OD} = \underline{d}$, $\overrightarrow{OC} = 2\underline{c}$ and $\overrightarrow{OE} = 3\underline{d}$

F is the midpoint of CD and $CG = \frac{1}{4}CE$.

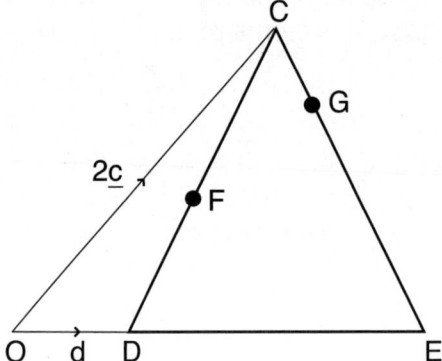

**(a)** Express in terms of $\underline{c}$ and $\underline{d}$.

**(i)** $\overrightarrow{DC}$ ........................................................................................ [1]

**(ii)** $\overrightarrow{DF}$ ........................................................................................ [1]

**(b)** Prove that O, F and G lie on a straight line.

........................................................................................

........................................................................................

........................................................................................ **[5]**

**(Total 7 marks)**

 **Examining Group**

**General Certificate of Secondary Education**

# Mathematics
# Higher Tier
# Exam 2    Paper 2

## Time: two hours

### Instructions to candidates

You **are expected to** use a calculator.

Write your name, centre number and candidate number in the boxes at the top of this page.

Answer ALL questions in the spaces provided on the question paper.

Show all stages in any calculations and state the units.

Include diagrams in your answers where this may be helpful.

### Information for candidates

The number of marks available is given in brackets **[2]** at the end of each question or part question.

The marks allocated and the spaces provided for your answers are a good indication of the length of answer required.

© 2003 Letts Educational

**EDUCATIONAL**

## Useful Formulae

**Volume of prism** = (area of cross section) × length

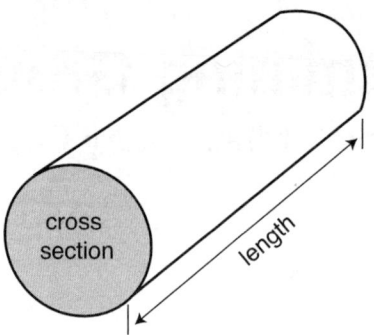

**In any triangle *ABC***

**Sine rule** $\dfrac{a}{\sin A} = \dfrac{b}{\sin B} = \dfrac{c}{\sin C}$

**Cosine rule** $a^2 = b^2 + c^2 - 2bc\cos A$

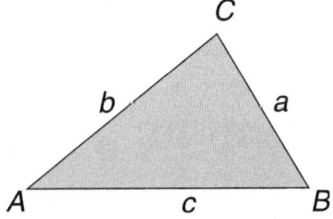

**Area of triangle** = $\dfrac{1}{2}ab\sin C$

**Volume of sphere** = $\dfrac{4}{3}\pi r^3$

**Surface area of sphere** = $4\pi r^2$

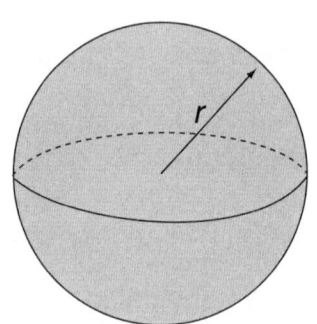

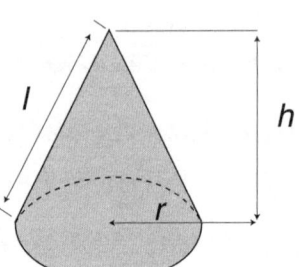

**Volume of cone** = $\dfrac{1}{3}\pi r^2 h$

**Curved area of cone** = $\pi r l$

**The quadratic equation**

The solutions of $ax^2 + bx + c = 0$, where a ≠ 0, are given by $x = \dfrac{-b \pm \sqrt{(b^2 - 4ac)}}{2a}$

*Letts*

**1**   Calculate:

$$\frac{2.76 \times 3.27^2 - 2.93\cos 30°}{\sqrt{12.62}\ 0.327}$$

Give your answer to **four** significant figures.

..............................................................................................

.......................................................................... **[3]**

**(Total 3 marks)**

**2**   $p$ is an integer such that $-4 \leqslant 4p < 14$.

**(a)**  List all the possible values of $p$.

.......................................................................... **[2]**

**(b)**  Solve the inequality:

$$\frac{3t + 2}{4} < t - 3$$

.......................................................................... **[3]**

**(Total 5 marks)**

3    (a)  Express the following numbers as products of their prime factors.

   (i)  40

   .......................................................................................................... [2]

   (ii)  105

   .......................................................................................................... [2]

   (b)  Find the Highest Common Factor of 40 and 105.

   .......................................................................................................... [1]

   (c)  Work out the Lowest Common Multiple of 40 and 105.

   .......................................................................................................... [2]

   (d)  Change the decimal  $0.3\dot{6}$  into a fraction in its lowest terms.

   .......................................................................................................... [3]

                                                                    **(Total 10 marks)**

4    The diagram represents a triangular playground DEF.
     The scale of the diagram is: 1cm represents 2m.
     A slide is to be placed in the playground so that it is:

     12m from point **F**
     equidistant between **D** and **E**.

     On the diagram, mark the point with a letter **S** where the slide can be placed.

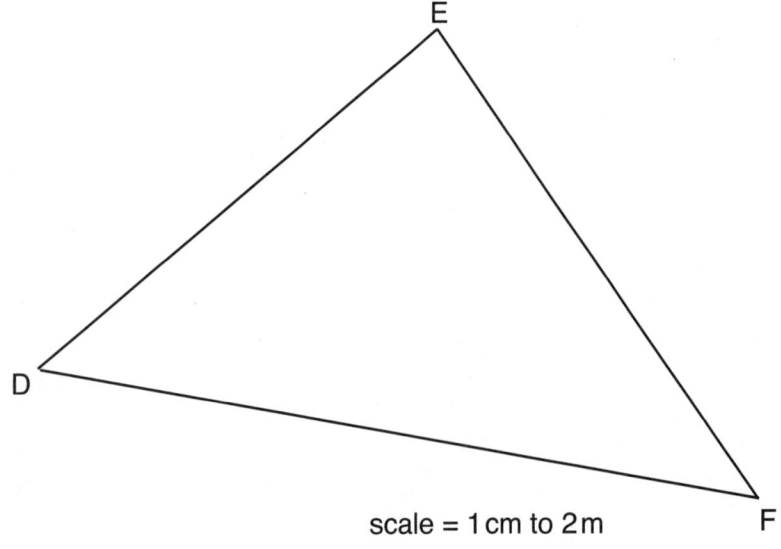

scale = 1 cm to 2 m

                                                                    [3]
                                                                    **(Total 3 marks)**

5   Use the method of trial and improvement to solve the equation:

$t^3 + 2t = 39$

Give your answer to one decimal place.
You must show all your working.

..............................................................................................................

..............................................................................................................

..............................................................................................................

$t$  .......................... [4]
**(Total 4 marks)**

6   The grouped frequency table shows information about the number of hours spent travelling by each of 60 commuters in one week.

| Number of hours spent travelling ($t$) | Frequency |
|---|---|
| $0 < t \leqslant 5$ | 0 |
| $5 < t \leqslant 10$ | 14 |
| $10 < t \leqslant 15$ | 21 |
| $15 < t \leqslant 20$ | 15 |
| $20 < t \leqslant 25$ | 7 |
| $25 < t \leqslant 30$ | 3 |

(a)  Find the class interval in which the median lies.

..............................................................................................................

.......................................................................................................... [2]

(b)  Work out an estimate for the mean number of hours spent travelling by commuters that week.

..............................................................................................................

..............................................................................................................

..............................................................................................................

.......................................................................................................... [4]
**(Total 6 marks)**

*Letts*

**7**

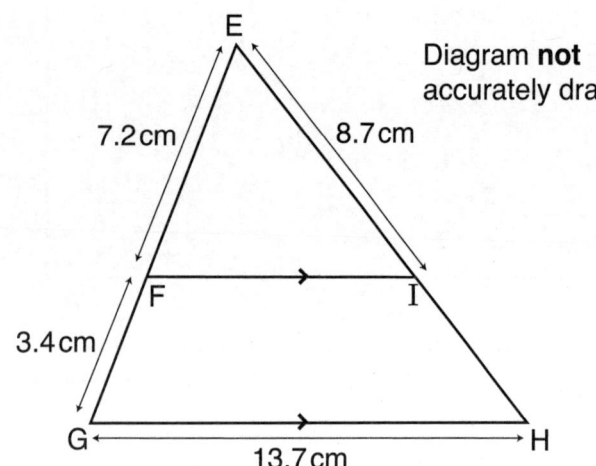

Diagram **not**
accurately drawn

FI is parallel to GH

EF = 7.2 cm, FG = 3.4 cm

EI = 8.7 cm, GH = 13.7 cm

**(a)** Calculate the length of IH.

........................................................................................................

........................................................................................................

........................................................................................................

.......................... cm    **[3]**

**(b)** Calculate the length of FI.

........................................................................................................

........................................................................................................

........................................................................................................

.......................... cm    **[2]**
**(Total 5 marks)**

**8**

$$p^3 = \frac{a^2b}{a \times b}$$

$a = 4 \times 10^6$

$b = 2 \times 10^3$

Find $p$.

Give your answer in standard form correct to 1 significant figure.

..............................................................................................................

..............................................................................................................

..............................................................................................................

$p = $ ................................ **[3]**
**(Total 3 marks)**

**[turn over**

**9**

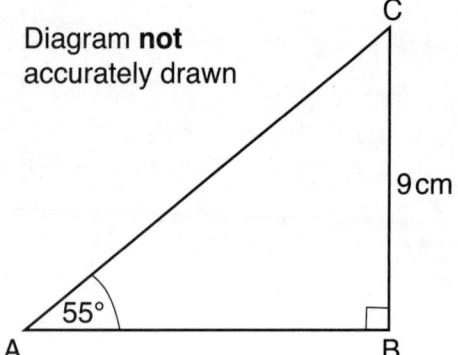

Diagram **not** accurately drawn

The diagram shows triangle ABC.

BC = 9cm
Angle ABC = 90°
Angle CAB = 55°

Work out the perimeter of the triangle.
Give your answer correct to 3 significant figures.

.............................................................................................................

.............................................................................................................

.............................................................................................................

.............................................................................................................

.............. cm    **[5]**
**(Total 5 marks)**

**10** A shop sells a television set.
It offers a discount of 15% off the normal price.
Ahmed buys the television set for £357.
Calculate the normal price of the television set.

.............................................................................................

.............................................................................................

.............................................................................................

£ ............................. **[3]**
**(Total 3 marks)**

**11**

|       | Year 7 | Year 10 |
|-------|--------|---------|
| Boys  | 90     | 50      |
| Girls | 85     | 70      |

The table shows the number of boys and the number of girls in year 7 and year 10.
The deputy head wants to find out how much homework pupils have per week.
A stratified sample of size 50 is to be taken from year 7 and year 10.

**(a)** Calculate the number of pupils to be sampled from year 7.

.............................................................................................

.............................................................................................

.............................................................................................

............................. **[2]**

**(b)** Two pupils are to be chosen at random to speak to the headteacher.
One pupil is to be chosen from year 7.
One pupil is to be chosen from year 10.

Calculate the probability that a girl and a boy will be chosen.

.............................................................................................

.............................................................................................

.............................................................................................

............................. **[3]**
**(Total 5 marks)**

**12** This is the graph of $y = 2x^2 - 3x + 2$.

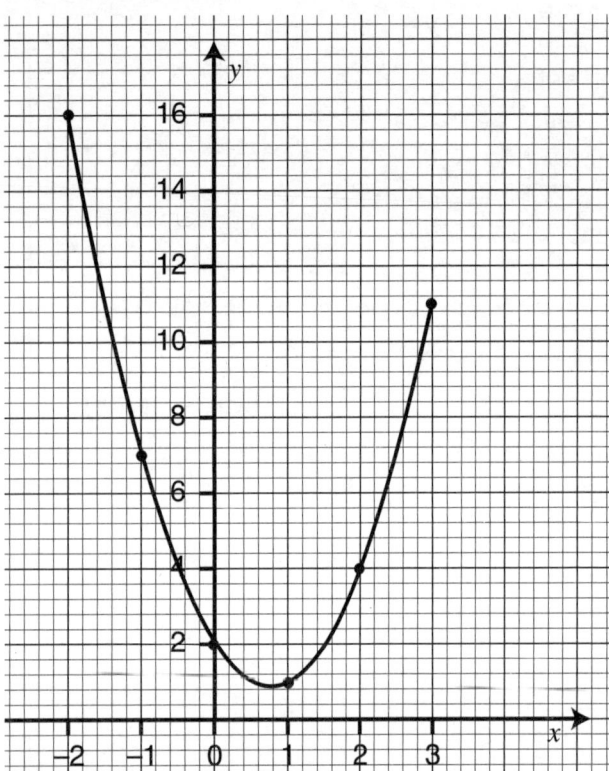

Use the graph to solve the following equations.

**(a)** $2x^2 - 3x + 2 = 2$

........................... **[2]**

**(b)** $2x^2 - 3x + 2 = 10$

........................... **[2]**

**(c)** $2x^2 - 4x - 2 = 0$

........................... **[4]**

**(Total 8 marks)**

**13** The length of a rectangle is 6.2 cm, correct to 1 decimal place.
The width of the rectangle is 3.76 cm correct to 2 decimal places.

**(a)** Calculate the upper bound for the area of the rectangle.
Write down all the figures on your calculator display.

..................................................................................................

..................................................................................................

..................................................................................................

........................ cm² **[3]**

**(b)** $a = 7.46$ cm, correct to 2 decimal places.
$b = 6.3$ cm, correct to 1 decimal place.

Calculate the lower bound for $\dfrac{a^2}{b}$ .

Write down all the figures on your calculator display.

..................................................................................................

..................................................................................................

..................................................................................................

........................ cm **[3]**
**(Total 6 marks)**

**14**

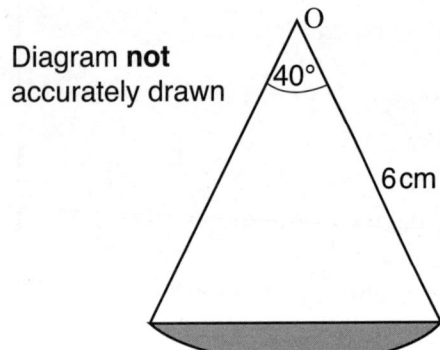

Diagram **not** accurately drawn

The diagram shows a sector of a circle centre O.
The radius of the circle is 6 cm.
The angle at the centre of the circle is 40°.

Calculate the area of shaded segment.
Give your answer correct to 3 decimal places.

..........................................................................................................

..........................................................................................................

..........................................................................................................

..........................................................................................................

..........................................................................................................

..........................................................................................................

....................... cm$^2$   **[6]**
**(Total 6 marks)**

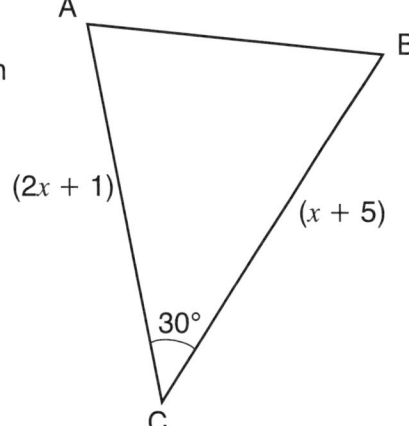

Diagram **not** accurately drawn

The diagram shows a triangle.
The measurements of the diagram are in centimetres.
The lengths of the two sides are $(2x + 1)$ cm and $(x + 5)$ cm.
Angle ACB is 30°.
The area of the triangle is 10 cm².

**(a)** Show that:

$2x^2 + 11x - 35 = 0$

...................................................................................................

...................................................................................................

...................................................................................................

...................................................................................................

................................................................................................... **[4]**

**(b)** Find the value of $x$.
Give your answer correct to 2 decimal places.

...................................................................................................

...................................................................................................

...................................................................................................

...................................................................................................

................................................................................................... **[3]**

**(Total 7 marks)**

**16** The population of a country is increasing at an annual rate of 1.3%.
In 1992, it was 42 million.
Calculate an estimate for the population of this country in 2012.

......................................................................................................................

......................................................................................................................

......................................................................................................................

.................................................................................................... million    **[3]**

**(Total 3 marks)**

**17** Solve the simultaneous equations:

$y = 2 - 3x$
$x^2 + y^2 = 58$

......................................................................................................................

......................................................................................................................

......................................................................................................................

....................................................................................................    **[6]**

**(Total 6 marks)**

**18** The diagram shows a quadrilateral ABCD.

Diagram **not**
accurately drawn

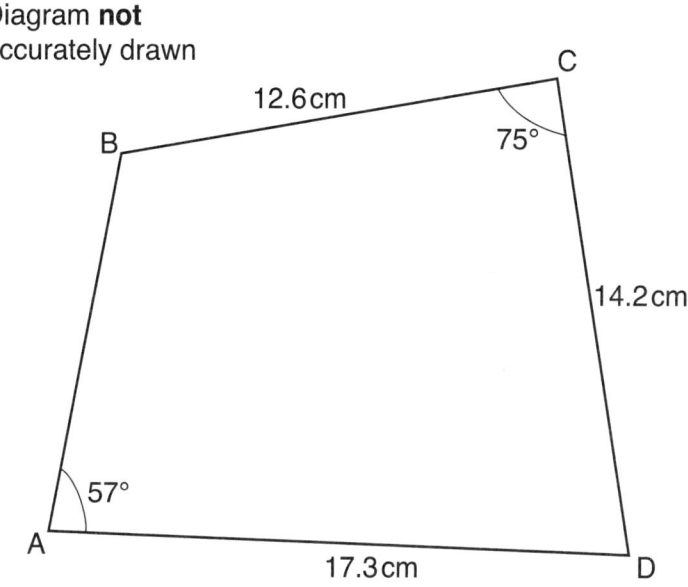

AD = 17.3 cm, BC = 12.6 cm, CD = 14.2 cm
Angle BCD = 75°, Angle BAD = 57°.

Calculate the size of angle ABD.
Give your answer correct to 2 decimal places.

.................................................................................................................

.................................................................................................................

.................................................................................................................

.................................................................................................................

.................................................................................................................

.................................................................................................................

...............° **[6]**
**(Total 6 marks)**

**19 (a)** Prove algebraically that the sum of the squares of any two consecutive integers is an odd number.

...............................................................................................................

...............................................................................................................

............................................................................................................... **[3]**

**(b)** Show that $(2n + 1)^2 - (n + 2)^2 = 3(n - 1)(n + 1)$

...............................................................................................................

...............................................................................................................

............................................................................................................... **[3]**

**(Total 6 marks)**

# Letts Examining Group

### General Certificate of Secondary Education

## Mathematics
## Higher Tier

# Mark scheme
# and
# Examiner's report

| Question | Answer | Mark |
|---|---|---|

**1 a i** 7.752    **1**
   **ii** 0.057    **1**

  **b**   $2 \times 3 \times 5 \times 7$    **2**

**Examiner's Tip**

In part (a), check that the sizes of your answers are reasonable. You will score 1 mark if you find two prime factors in part (b).

**2 a**   $7x - 4x + 12 = 27$    **1**
      $3x = 15$    **1**
      $x = 5$    **1**

  **b**   $5(x + 3) - 2(x - 4) = 50$    **1**
      $5x + 15 - 2x + 8 = 50$    **1**
      $3x = 27$    **1**
      $x = 9$    **1**

**Examiner's Tip**

You can combine steps and still earn the marks but take care not to make a mistake. The negative sign in front of the fraction could catch you out.

**3**   $\dfrac{36 \times 5}{48} \times 10^{-5}$    **1**

    $\dfrac{15}{4} \times 10^{-5}$    **1**

    $3.75 \times 10^{-5}$    **1**

**Examiner's Tip**

There is 1 mark for simplifying the powers of 10 and 1 mark for the other numbers. 0.0000375 will score 2 marks.

**4**   *Perpendicular bisector of AB*    **1**
    *Angle bisector of angle CAB*    **1**
    *Circle centre A radius 6 cm*    **1**
    *Shading as shown*    **1**

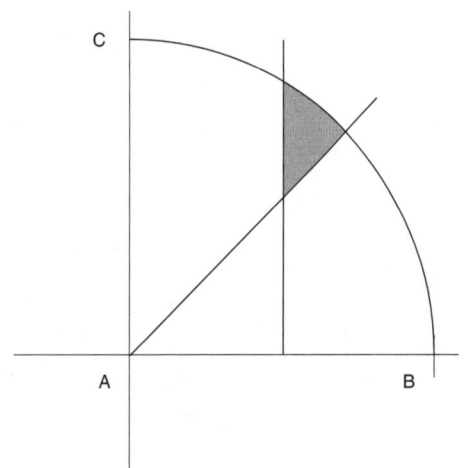

**5 a**   *Curve passing through $(-2, 1)$, $(-1, 5)$, $(0, 3)$, $(1, 1)$, $(2, 5)$*    **4**

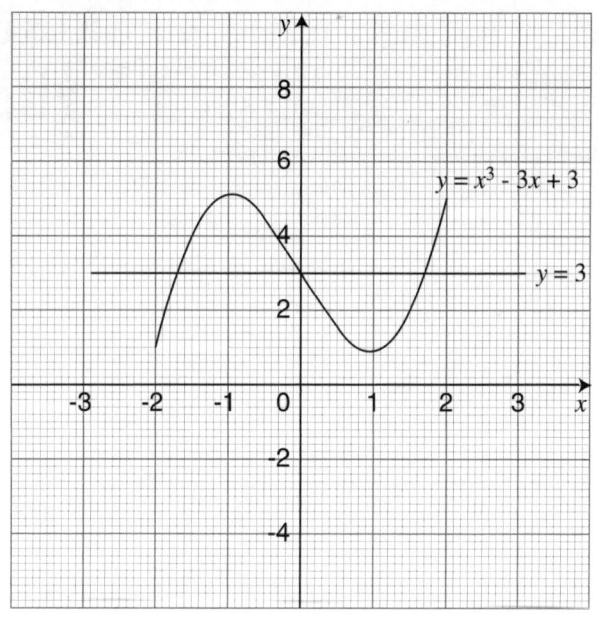

  **b i**   $y = 3$ drawn    **1**
       1.7(3), 0, $-1.7(3)$    **1**

    **ii**   $x^3 - 3x + 3 = 3$
        or $x^3 - 3x = 0$    **1**

**6**   Angle ABO = $90 - x$ (Radius perpendicular to tangent)    **1**
    Angle BAO = $90 - x$ (Isosceles triangle)    **1**
    Angle AOB = $180 - 2(90 - x)$ (Angles of a triangle)    **1**
    $= 2x$    **1**

**7 a**   triangle: $0 = 9A + 3B$   or $3A + B = 0$    **1**
       quadrilateral: $2 = 16A + 4B$   or $8A + 2B = 1$    **1**

  **b**   $8A + 2B = 1$
      $\underline{6A + 2B = 0}$   (subtract)    **1**
      $2A \quad\quad = 1 \quad\quad A = \frac{1}{2}$    **1**
      $B = -1\frac{1}{2}$    **1**

  **c**   $d = \frac{1}{2} \times 15^2 - 1\frac{1}{2} \times 15 = 90$    **1**

**Examiner's Tip**

You may have used a different method for solving the equations, substituting $B = -3A$ in $8A + 2B = 1$ for instance. All correct algebraic methods are acceptable but not trial and improvement (or trial and error!) in such a case. The clue to this was in the question where the instruction was to 'Use algebra...'.

| Question | Answer | Mark |
|---|---|---|

**8 a** Angle $BAC = 80°$ and angle $EDF = 70°$
so the triangles are equiangular
and therefore similar. **1**

**b** $\dfrac{AB}{AC} = \dfrac{DF}{EF}$ **1**

$AB = 12 \times \frac{2}{4} = 6\,\text{cm}$ **1**

$\dfrac{DE}{EF} = \dfrac{BC}{AC}$ **1**

$DE = 4 \times \frac{9}{12} = 3\,\text{cm}$ **1**

**Examiner's Tip**

Corresponding sides are opposite equal angles, e.g. AB corresponds to DF, both opposite 30°.

---

**9** Ratio of Education : Other Services = 7 : 3 **1**
Amount spent on other services = $420 \times \frac{3}{7}$ **1**
$= £180$ million **1**

**10a** $r = 600$ **1**
$s = 200$ **1**

**b** 800, 1250, 2200, 2800 **1**

**c** *points plotted at right hand end of intervals* **1**
*points joined with a smooth curve or ruled line*
*segments* **1**

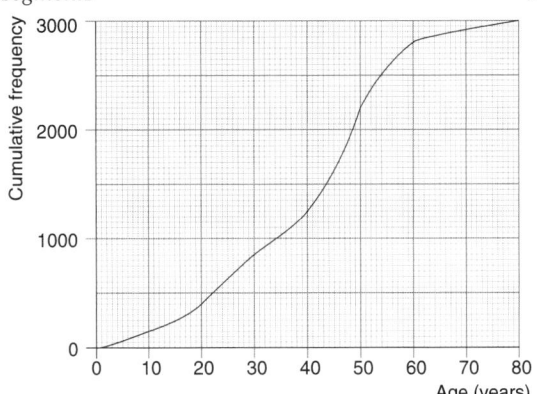

**d i** about 380 people **2**

**ii** about 43 years **1**

**Examiner's Tip**

Remember that it is the *area* under the histogram that gives the frequency. This means multiplying the column width by the height. For s, this is $10 \times 20$. A cumulative frequency graph can be a smooth curve or a series of straight lines. However, you should draw a curve if that is what it says in the question. For part (d)(i), read the cumulative number of people at 57 years (for 1 mark) and then subtract this from 3000. For the median, read the age that corresponds to 1500 cumulative frequency, half the total.

---

| Question | Answer | Mark |
|---|---|---|

**11** There are three different relationships to consider:
$18 + (4 + 3x) > 6 - x$
i.e. $22 + 3x > 6 - x$
i.e. $4x > -16$
$x > -4$ **1**

$18 + (6 - x) > 4 + 3x$
$24 - x > 4 + 3x$
$20 > 4x$
$5 > x$
i.e. $x < 5$ **1**

$(6 - x) + (4 + 3x) > 18$
$10 + 2x > 18$
$2x > 8$
$x > 4$ **1**

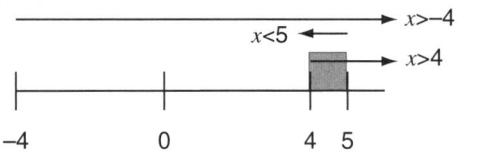

solution is $4 < x < 5$ **2**

**12a** angle CBD **1**
angle CED **1**

**b** $x =$ angle BDE = angle BCE (subtended by the same arc BE) **1**
angle BCE $+ y +$ angle BEC $= 180°$ (angles of a triangle) **1**
$x + y = 90°$ (angle BEC $= 90°$, angle in a semicircle) **1**

**c** ABC, BDC, ADB **2**

**13** $x^2 + (x - 7)^2 = 25$ **1**
$2x^2 - 14x + 24 = 0$ **1**
$(x - 3)(x - 4) = 0$ **1**
$x = 3, 4$ **1**
$y = -4, -3$ **1**
Points of intersection (3, -4) and (4, -3) **1**

**14a** $(2x + 1)(x - 3)$ **1 + 1**
$x = -\frac{1}{2}$ or 3 **1**

**b** $6y - 5y^2 + y^3 + y^3 + 5y^2$ **1**
$2y^3 + 6y$ **1**
$2y(y^2 + 3)$ **1**

**c** $8gh = 3u^2$ **1**
$u^2 = \frac{8}{3}gh$ **1**
$u = \pm\sqrt{\frac{8}{3}gh}$ **1**

**Examiner's Tip**

You could take out a factor of $y$ in part (b) before expanding the bracket,
i.e. $y(6 - 5y + y^2 + y^2 + 5y)$.

---

| Question | Answer | Mark |
|---|---|---|
| **15a** | $\dfrac{3}{5}$ | 1 |
| | $\times\dfrac{2}{4}$ | 1 |
| | $=\dfrac{3}{10}$ | 1 |
| **b** | $p(BBP) = \dfrac{3}{10}\times\dfrac{2}{3} = \dfrac{1}{5}$ | 1 + 1 |
| | $p(BP \text{ or } PB) = 2\times\dfrac{3}{5}\times\dfrac{2}{4} = \dfrac{6}{10}$ | 1 |
| | $p(BPP \text{ or } PBP) = \dfrac{6}{10}\times\dfrac{1}{3} = \dfrac{1}{5}$ | 1 |
| | $p(\text{third tin P}) = \dfrac{1}{5}+\dfrac{1}{5} = \dfrac{2}{5}$ | 1 |

| Question | Answer | Mark |
|---|---|---|
| **16a** | $\overrightarrow{AB} = b - a$ | 1 |
| | $\overrightarrow{PB} = \tfrac{3}{4}(b - a)$ | 1 |
| | $\overrightarrow{OP} = \overrightarrow{OA} + \overrightarrow{AP} = \tfrac{1}{4}(3a + b)$ | 1 + 1 |
| **b** | $\overrightarrow{BQ} = \overrightarrow{OQ} - \overrightarrow{OB}$ | 1 |
| | $= \overrightarrow{OA} + \overrightarrow{AQ} - \overrightarrow{OB}$ | 1 |
| | $= \overrightarrow{OA} + \overrightarrow{OP} - \overrightarrow{OB}$ | 1 |
| | $= a + \tfrac{1}{4}(3a + b) - b$ | 1 |
| | $= \tfrac{1}{4}(7a - 3b)$ | 1 |
| **17a** | $100a + 10b + c$ | 2 |
| **b** | $100a + 10b + c - (100c + 10b + a)$ | 1 |
| | $100a - 100c + c - a$ | 1 |
| | $99a - 99c$, which has a factor of 9 | 1 |

**Examiner's Tip**
The algebra is very simple once you have allowed for the place value.

**1**    *Calculations*: number of books has increased by nearly 13 times;
number of readers has increased by over 5 times;
number of borrowings has increased by nearly 10 times.    **1**
*Conclusions*: each reader reads more books $(10 > 5)$;    **1**
each book is used less often $(13 > 10)$    **1**

**Examiner's Tip**
You may have done the calculations differently or drawn different conclusions. However, so long as you have said more than, for instance, 'The number of books has increased', equivalent work will earn the marks. The important thing is to combine some of the data.

**2 a**

|  | Median | Interquartile range | Tallest height |
|---|---|---|---|
| Girls | 1.30 | 0.20 | 1.54 |
| Boys | 1.36 | 0.23 | 1.78 |

   **4**

**b**    Two from
Boys are taller on average
The tallest is a boy
The shortest boy is the same height as the shortest girl
The boy's heights are more varied    **2**

**3 a**    $x = -2, 0.8, 3.1$    **1 + 1**

**b**    $x = 3.1$   expression $= 0.071$    **1**
$x = 3.05$   expression $= -0.482375$
$x = 3.08$   expression $= -0.154688$    **1**
$x = 3.09$   expression $= -0.042571$
$x = 3.095$   expression $= 0.0140$    **1**
$x = 3.09$ to 2 decimal places    **1**

**Examiner's Tip**
The last two calculations show that the root is between 3.09 and 3.095 as it is between these that the sign changes. If you do not show that the root is nearer 3.09 than 3.10, you will lose a mark.

**4**

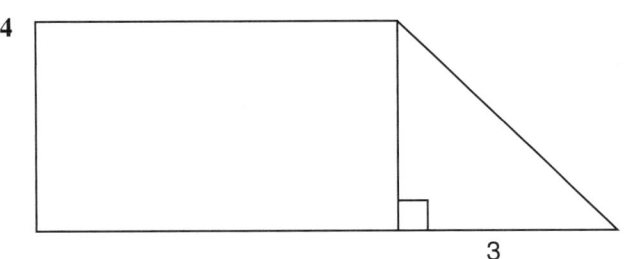

identifying right angled triangle    **1**
using $8 - 5 = 3$    **1**

distance between parallels $= 3\tan40°$    **1**
area $= \frac{1}{2} \times 3\tan40° \times (5 + 8)$    **1**

---

$= 16.4\text{cm}^2$    **1**

**5**    number of minutes $= 16 \times 365 \times 24 \times 60$    **1**
$2 \div (16 \times 365 \times 24 \times 60)$    **1**
$= 2.37823... \times 10^{-7}$    **1**
$= 2.4 \times 10^{-7}$ or $2.38 \times 10^{-7}$    **1**

**Examiner's Tip**
For this sort of calculation it is not necessary to include leap years, as the answer would be the same to an appropriate degree of accuracy − try the calculations again using 365.25 instead to check that this is true. 0.00000024 would earn 3 marks.

**6 a**    gradient $= \dfrac{13 - 5}{6 - 2}$    **1**
$= 2$    **1**
intercept $= 1$    **1**
equation is $y = 2x + 1$    **1**

**b**    lines meet when $4 - x = 2x + 1$    **1**
$x = 1, y = 3$    **1**
C is at $(0, 4)$    **1**
area $= \frac{1}{2} \times 1 \times 3 = 1\frac{1}{2}$    **1**

**Examiner's Tip**
To find the intercept in part (a), an easy way is to work left from $(2,5)$ with gradient 2. 2 left for $x$ means $2 \times 2$ down for $y$.

**7 a**

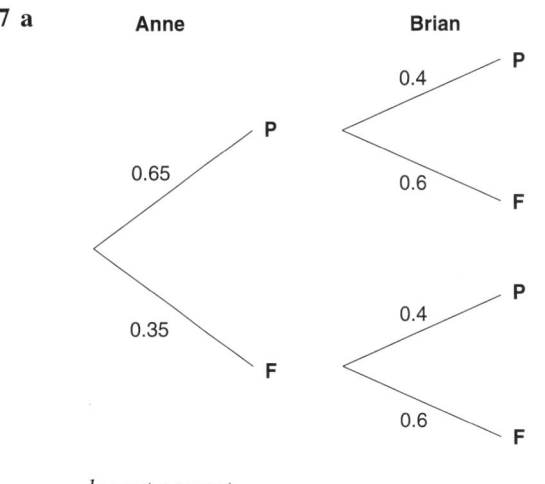

layout correct    **1**
Anne correct    **1**
Brian correct    **1**

**b i**    $0.65 \times 0.4 = 0.26$    **1**

**ii**    $0.65 \times 0.6$    **1**
$+ 0.35 \times 0.4$    **1**
$= 0.53$    **1**

**Examiner's Tip**
Be sure to set out your working clearly. There are other ways of labelling a tree diagram which are just as good, but be consistent!

| Question Answer | Mark |
|---|---|

**8 a** $r^2 + y^2 = (r + x)^2$    **1**
$r^2 + y^2 = r^2 + 2rx + x^2$    **1**

$y^2 = 2rx + x^2$   or $y = \sqrt{2rx + x^2}$    **1**

**b** $150^2 = 2 \times 6340 \times x + x^2$    **1**
$x^2 + 12680x - 22500 = 0$    **1**

$x = \dfrac{-12680 + \sqrt{12680^2 + 4 \times 22500}}{2}$    **1**

$= 1.77$    **1**

**c i** If $x$ is small compared with $r$, $x^2$ is much smaller than $2rx$ and can be ignored in an approximate formula.    **1**

**ii** $y = \sqrt{2 \times 6340 \times 1.2}$    **1**
$= 123\,\text{km}$    **1**

**9 a** $12000 \times 1.065^5$    **1**
$= £16441$    **1**

**b** $16441 + 3000$    **1**
1 year   $19441 \times 1.065 =$    $20704$
2 years   22050    **1**
3 years   23483
4 years   25010    **1**
greater than £24000 after 4 years    **1**

**Examiner's Tip**
You don't need to write down the intermediate answers in part (b) but be sure to check the calculations. If you interpreted the time in part (b) to include the first year, an answer of 5 will be acceptable.

**10a** area of base $= 125 \div 12$    **1**

radius $= \sqrt{\dfrac{h}{\pi}} = \sqrt{\dfrac{125}{12\pi}}$    **1**

$= 1.82$    **1**

**b** $125 = \dfrac{\pi}{3} \times h \times \left(\dfrac{h}{2}\right)^2$    **1**

$h^3 = \dfrac{12 \times 125}{\pi}$    **1**

$h = 7.82\,\text{cm}$    **1**

**c** volume scale factor $= \left(\dfrac{1}{2}\right)^3 = \dfrac{1}{8}$    **1**

8 can be filled    **1**

**Examiner's Tip**
Don't forget that the formula for the volume of a cone is on page 2 of the question paper.

| Question Answer | Mark |
|---|---|

**11** $ART = 100(P - A)$    **1**
$= 100P - 100A$    **1**
$A(RT + 100) = 100P$    **1**

$A = \dfrac{100P}{RT + 100}$    **1**

**Examiner's Tip**
To avoid errors, do one step at a time!

**12**

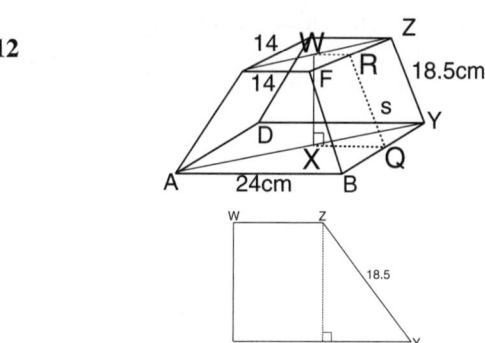

**a i** consider slice WXYZ
WZ $= 0.5 \times$ diagonal of top $=$
$0.5 \times \sqrt{14^2 + 14^2}$

XY $= 0.5 \times$ diagonal of base $=$
$0.5 \times \sqrt{24^2 + 24^2}$    **1**

PY $=$ XY $-$ YZ
$= 16.971 - 9.899 = 7.071$    **1**
PZ$^2 = 18.5^2 - 7.071^2$    **1**
therefore PZ $= 17.1\,\text{cm}$    **1**

*answer only = 4 marks*

**ii** in slice WXQR

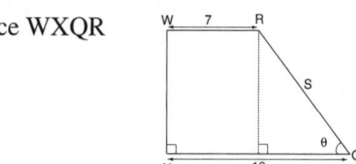

$s^2 = 17.1^2 + (12\text{-}7)^2$    **1**
$s = 17.8\,\text{cm}$    **1**

*answer only = 2 marks*

**iii** $\tan\theta = \dfrac{17.1}{5}$    **1**

$\theta = 73.7°$    **1**

*answer only = 2 marks*

**iv** Area $= 4\left(\dfrac{1}{2}(14 + 24) \times 17.8\right) + 14^2 + 24^2$    **1**

$= 2124.8\,\text{cm}^2$    **1**

*answer only = 2 marks*

| Question | Answer | Mark |
|---|---|---|
| **13** | $3(2x + 1) - (x - 1) = (x - 1)(2x + 1)$ | 1 |
| | $6x + 3 - x + 1 = 2x^2 - x - 1$ | 1 |
| | $2x^2 - 6x - 5 = 0$ | 1 |
| | $x = \dfrac{6 \pm \sqrt{36 + 40}}{4}$ | 1 |
| | $x = 3.68$ or $-0.68$ | 2 |
| **14a** | Sales fluctuate across four quarters. | 1 |
| **b** | Moving averages: 17, 17.5, 17.5, 17, 16 | 2 |
| | Plotted in middle of four relevant points | 1 |
| **c** | $15.5 \times 4 - 15 - 25 - 12 = 10$ | 1 |
| **15** | minimum safe load = 14.5 tonnes | 1 |
| | maximum block size = $0.305 \times 0.105 \times 0.105$ | 1 |
| | maximum density = 1550 | 1 |
| | $14500 \div (0.305 \times 0.105^2 \times 1550)$ | 1 |
| | = 2782 | 1 |

**Examiner's Tip**

For safety, assume the minimum for the capacity of the truck and the maximum for the mass of the blocks.

| Question | Answer | Mark |
|---|---|---|
| **16a** | Numerator = $(x - 3)(x + 3)$ | 1 |
| | Denominator = $(3x + 2)(x - 3)$ | 1 + 1 |
| | $= \dfrac{x + 3}{3x + 2}$ | 1 |
| **b** | Use of $x(x + 4)$ for denominator | 1 |
| | $\dfrac{3x + 4(x + 4)}{x(x + 4)}$ | 1 |
| | $\dfrac{7x + 16}{x(x + 4)}$ | 1 |

**Examiner's Tip**

The rules for fractions in algebra are just the same as those in arithmetic.

| Question | Answer | | Mark | Question | Answer | | Mark |
|---|---|---|---|---|---|---|---|

**1**

$$\frac{60 \times 5 - 10 \times 3}{100^2}$$

$$= \frac{270}{10000}$$

$$= \frac{27}{1000}$$

**3**

**Examiner's tip**
Round each value to 1 significant figure.

**2**     $3n + 4$                                 **2**

**3**   **i**   898.8                              **1**
      **ii**   8.988                              **1**
     **iii**   32.1                               **1**

**4**  **a**   Reflection in the line $x = \dfrac{1}{2}$   **1**

      **b**                                        **2**

      **c**                                        **2**

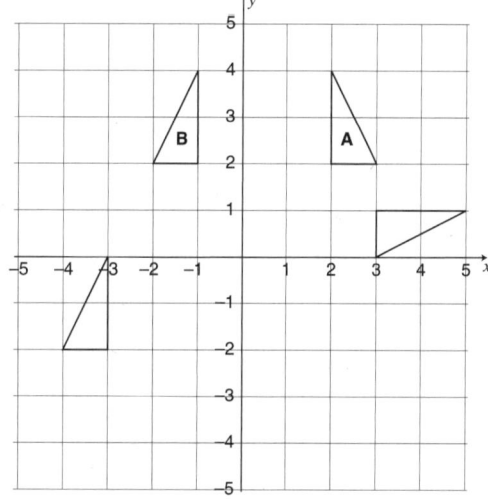

**5**  **a**   $a = -4$                            **2**

**Examiner's tip**
Remember to multiply out the brackets first and then collect like terms.

**6**      $a^2 + bc$        Area

          $\dfrac{abc}{\pi b^2}$        Length

          $3a^2\sqrt{b^2 + c^2}$        Volume        **3**

**7**  **a**  **i**   lower quartile: 2 minutes
          **ii**   interquartile range: 3 minutes    **3**

      **b**
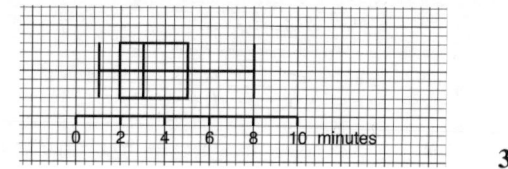
                                                   **3**

**Examiner's tip**
Remember that each section of the box plot represents 25%.

**8**  **a**   $1.8 \times 10^{12}$               **2**

      **b**
$$\frac{(6 \times 10^7)^2}{3 \times 10^4} = \frac{36 \times 10^{14}}{3 \times 10^4}$$

$$= 12 \times 10^{10}$$

$$= 1.2 \times 10^{11}$$

                                                   **3**

**Examiner's tip**
You can use the law of indices on these questions.
Remember to write your final answer in standard form.

**9**  **a**  **i**   $n^4$                        **1**
          **ii**   $\dfrac{2}{3h}$                 **1**

      **b**  **i**   $10x^2 + 29x - 21$            **2**
          **ii**   $25x^2 - 40x + 16$             **3**

      **c**      $x = 2$ or $x = -6$              **3**

**Examiner's tip**
In (b)(ii) remember that $(5x - 4)^2 = (5x - 4)(5x - 4)$. In part (c) when solving the quadratic equation, factorise first.

**10**  **i**   36°
           Since angle FEG = 90°, angles in a semicircle, then angle FGE = 180° − 90° − 54° since angles in a triangle add up to 180°.      **2**

       **ii**   54°
           Since angle NGO = 90° because a radius and tangent meet at 90° and FGE = 36°, and EGN = 54°.      **2**

**Examiner's tip**
When asked to 'explain' make sure you refer to the circle theorems and not just show working out.

*Letts*

**11 a**     $y = -2x + 8$     2

**b**     $-\dfrac{1}{3}$     2

**c**     $y = -2x - 1$     2

**d**     $y = \dfrac{-1}{3}x + \text{any constant}$     2

**Examiner's tip**

For part (d) remember that two lines are perpendicular if the gradients multiply to give –1 ( $\dfrac{-1}{3} \times 3 = -1$ ).

**12   i**   $1$   1

**ii**   $\dfrac{1}{8^2} = \dfrac{1}{64}$   1

**iii**   $(\sqrt[3]{27})^2 = 9$   1

**iv**   $\dfrac{1}{25^{-\frac{1}{2}}} = 25^{\frac{1}{2}} = \sqrt{25} = \pm 5$   2

**13**     $y = \dfrac{a(x + b)}{x - c}$

$y(x - c) = ax + ab$
$yx - yc = ax + ab$
$yx - ax = yc + ab$
$x(y - a) = yc + ab$

$x = \dfrac{cy + ab}{(y - a)}$     4

**Examiner's tip**

Remember to show each step in your working.

**14**     $\dfrac{144}{729} = \dfrac{16}{81}$

$1 - \dfrac{16}{81} = \dfrac{65}{81}$     6

**Examiner's tip**

You could draw a tree diagram to help you or work out the probability of drawing all three colours (i.e. red, green, blue) and mulitply by 6 as there are six possible combinations, then subtract this from 1.

**15**

$\dfrac{(7 + \sqrt{5})(7 - \sqrt{5})}{\sqrt{80}}$

$= \dfrac{(49 - 5)}{\sqrt{80}}$

$= \dfrac{44}{\sqrt{16} \times \sqrt{5}}$

$= \dfrac{44}{4\sqrt{5}}$

$= \dfrac{11}{\sqrt{5}} \times \dfrac{\sqrt{5}}{\sqrt{5}}$

$= \dfrac{11\sqrt{5}}{5}$     4

**Examiner's tip**

To rationalise the denominator, multiply both the numerator and the denominator by $\sqrt{5}$.

**16 a**     $a = 3b^2$     3

**b**     27     1

**c**     $\pm 8$     2

**17 a**

| Time ($t$) in minutes | Frequency |
|---|---|
| $0 < t \leqslant 5$ | 10 |
| $5 < t \leqslant 15$ | 52 |
| $15 < t \leqslant 30$ | 48 |
| $30 < t \leqslant 50$ | 88 |
| $50 < t \leqslant 60$ | 24 |

2

**b**

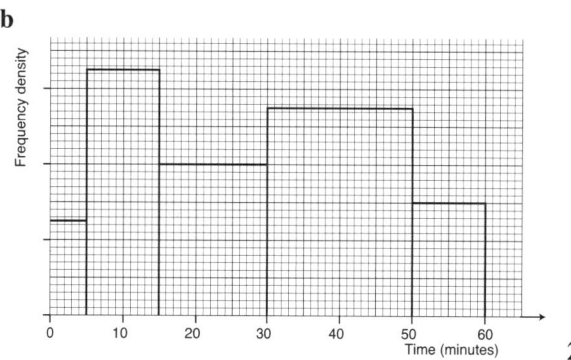

2

**18**

Cone: $\frac{1}{3}\pi r^2 h$

$= \frac{1}{3} \times \pi \times 9 \times 8$

$= 24\pi$

Hemisphere: $\frac{2}{3}\pi r^3$

$= \frac{2\pi \times 3^3}{3}$

$= 18\pi$

Volume $= 42\pi\,\text{cm}^3$      **7**

**19**

$\dfrac{2n^2 + n - 6}{4n^2 - 9} \times \dfrac{4n + 6}{n^2 + 3n + 2}$

$= \dfrac{(2n - 3)(n + 2)}{(2n - 3)(2n + 3)} \times \dfrac{2(2n + 3)}{(n + 1)(n + 2)}$

$= \dfrac{2}{(n + 1)}$      **6**

**Examiner's tip**
Factorise each part first and then cancel.

**20 a**    **i**    $\overrightarrow{DC} = -\underline{d} + 2\underline{c}$      **1**

       **ii**    $\overrightarrow{DF} = \frac{1}{2}(-\underline{d} + 2\underline{c})$      **1**

   **b**      $\overrightarrow{OF} = \frac{1}{2}(\underline{d} + 2\underline{c})$

         $\overrightarrow{OG} = \frac{3}{4}(\underline{d} + 2\underline{c})$      **5**

Since $\overrightarrow{OG}$ is a multiple of $\overrightarrow{OF}$

i.e. $\overrightarrow{OG} = \frac{3}{2}\overrightarrow{OF}$, and since they go

through the common point O; O, G and F must lie on a straight line.

**Examiner's tip**
Be very systematic when working out vectors.

| Question | Answer | Mark |
|---|---|---|
| **1** | 7.497 (4 s.f.) | **3** |

**Examiner's tip**
Make sure you show each step in your working.

| | | |
|---|---|---|
| **2 a** | $-1, 0, 1, 2, 3$ | **2** |
| **b** | $t > 14$ | **3** |

**Examiner's tip**
Solve inequalities in the same way that you would solve equations.

| | | | |
|---|---|---|---|
| **3 a** | **i** | $2 \times 2 \times 2 \times 5$ | **2** |
| | **ii** | $3 \times 5 \times 7$ | **2** |
| **b** | | HCF = 5 | **1** |
| **c** | | LCM = 840 | **2** |
| **d** | | $\frac{36}{99} = \frac{4}{11}$ | **3** |

**Examiner's tip**
When dividing to find factors, be systematic, e.g. start with 2. When no more factors of 2, try 3 and so on.

**4**

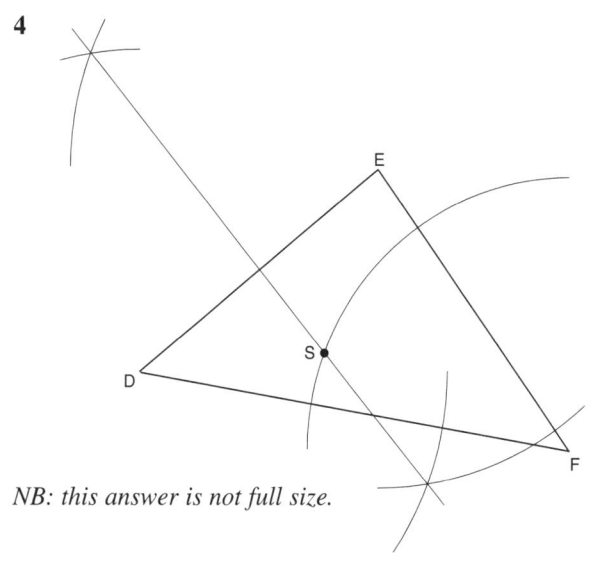

*NB: this answer is not full size.*

| | | |
|---|---|---|
| | | **3** |
| **5** | $t = 3.2$ | **4** |

**Examiner's tip**
It is important that you show full working out.

| | | |
|---|---|---|
| **6 a** | $10 < t \leqslant 15$ | **2** |
| **b** | 14.5 hours | **4** |

**Examiner's tip**
Remember to multiply the frequency with the midpoint in part (b) and then divide by 60 i.e. the sum of the frequency.

| Question | Answer | Mark |
|---|---|---|
| **7 a** | $12.8 - 8.7$ $= 4.1\,\text{cm}$ | **3** |
| **b** | $9.3\,\text{cm}$ | **2** |

**Examiner's tip**
In part (a) remember that the question asks for IH, so you need to subtract EH.

**8**

$$p = \sqrt[3]{\frac{(4 \times 10^6)^2 \times (2 \times 10^3)}{(4 \times 10^6) \times (2 \times 10^3)}}$$

$$= 2 \times 10^3 \qquad \textbf{3}$$

**9**
$$AB = \frac{9}{\tan 55°}$$
$$= 6.3018\ldots$$
$$AC = \frac{9}{\sin 55°}$$
$$= 10.98\ldots$$
$$\text{Perimeter} = 26.3\,\text{cm (3 s.f.)} \qquad \textbf{5}$$

**Examiner's tip**
To find AC, Pythagoras Theorem or trigonometry could have been used.

| | | |
|---|---|---|
| **10** | £420 | **3** |

**Examiner's tip**
Remember to divide by 0.85.

| | | |
|---|---|---|
| **11 a** | 30 year 7 pupils | **2** |
| **b** | $\left(\frac{85}{175} \times \frac{50}{120}\right) + \left(\frac{90}{175} \times \frac{70}{120}\right) = \frac{211}{420}$ | **3** |
| **12 a** | $x = 0$ or $x = \frac{3}{2}$ | **2** |
| **b** | $x = -1.39$ or $x = 2.89$ | **2** |
| **c** | $x = -0.41$ or $x = 2.41$ | **4** |
| **13 a** | $6.25 \times 3.765$ $= 25.53125\,\text{cm}^2$ | **3** |
| **b** | $\frac{7.455^2}{6.35}$ $= 8.752287402\,\text{cm}$ | **3** |

**14** Area of segment =
sector area − area of triangle

$$= \left(\frac{40°}{360°} \times \pi \times 6^2\right) - \left(\frac{1}{2} \times 6 \times 6 \times \sin 40°\right)$$

$$= 0.996\,\text{cm}^2 \;(3\text{ d.p.}) \qquad\qquad\qquad \mathbf{6}$$

**Examiner's tip**
Answer this question systematically and show all steps in your working.

**15 a** $\quad \frac{1}{2} \times (2x + 1) \times (x + 5) \times \sin 30° = 10$

$$(2x + 1)(x + 5) = 40 \quad (\sin 30° = \frac{1}{2})$$
$$2x^2 + 11x + 5 = 40$$
$$2x^2 + 11x - 35 = 0 \qquad\qquad\qquad \mathbf{4}$$

**b**
$$x = \frac{-b \pm \sqrt{b^2 - 4ac}}{2a}$$

$$x = \frac{-11 \pm \sqrt{11^2 - (4 \times 2 \times -35)}}{4}$$

$$x = \frac{-11 \pm \sqrt{121 + 280}}{4}$$

$$x = \frac{-11 \pm \sqrt{401}}{4}$$

$$x = \frac{-11 + 20.02}{4} \quad \text{or} \quad x = \frac{-11 - 20.02}{4}$$

$$x = 2.256 \qquad \text{or} \qquad x = -7.755$$

since $x$ represents a length:

$$x = 2.26\,\text{cm} \;(2\text{ d.p.}) \qquad\qquad\qquad \mathbf{3}$$

**Examiner's tip**
Even if you cannot do part (a), part (b) is simply solving a quadratic equation by using a formula.

**16** $\quad 42 \times (1.013)^{20}$
$\quad = 54.4$ million $\qquad\qquad\qquad \mathbf{3}$

**Examiner's tip**
The use of a multiplier is important when working out the answer. Do not attempt to work it out for each year!

**17** $\quad x = 3, y = -7$
$\quad x = -1.8, y = 7.4 \qquad\qquad\qquad \mathbf{6}$

**18** Calculate BD

$$BD = \sqrt{(12.6)^2 + (14.2)^2 - (2 \times 12.6 \times 14.2 \times \cos 75°)}$$

$$BD = 16.36\,\text{cm}$$

$$\frac{\sin A\hat{B}D}{17.3} = \frac{\sin 57°}{16.36}$$

$$\sin A\hat{B}D = \frac{\sin 57°}{16.36} \times 17.3$$

$$\text{angle ABD} = 62.48° \qquad\qquad\qquad \mathbf{6}$$

**Examiner's tip**
This is an example of a multistepped question. First find the length of BD by using the cosine rule and then the size of angle ABD by using the sine rule.

**19**   **a**   $n^2 + (n + 1)^2$
$n^2 + (n^2 + 2n + 1)$
$= 2n^2 + 2n + 1$
$\underbrace{2n(n + 1)} + 1$
Always even $\qquad\qquad\qquad \mathbf{3}$
$2n(n + 1)$ must always be even as it
is a multiple of 2, then $2n(n + 1) + 1$ will
always be odd.

  **b**   $(2n + 1)^2 - (n + 2)^2$
$(4n^2 + 4n + 1) - (n^2 + 4n + 4)$
$= 3n^2 - 3$
$= 3(n^2 - 1)$
$= 3(n - 1)(n + 1) \qquad\qquad \mathbf{3}$

# Examiner's report and grade predictor

## WHAT IS IN THE EXAM?

The knowledge and skills which the papers are designed to assess are from the Programmes of Study in the National Curriculum. The papers reflect the changes in the National Curriculum and also the recommendations in reports produced for the Government. These include asking more questions on algebra and some questions involving several steps without any prompts being provided. Half of the examination must be done without a calculator.

Most syllabuses divide the examination into three parts. Twenty percent of the marks are given for coursework, marked by your teachers or the examination board. The coursework consists of two tasks: one on Using and Applying Mathematics and one on Data Handling. The remaining eighty percent is tested by two examination papers, each usually marked out of 100. On each of these papers, the questions are set on Number, Algebra, Shape, Space and Measures and Data Handling.

In Algebra, most of the marks are given for solving equations, changing formulas and other algebraic manipulation. Fewer marks will be given for graphical work and sequences. In Shape and Space, some of the formulas you need are included on the formula sheet, usually at the front of the question paper. These include areas, volumes and trigonometry.

## WHICH QUESTIONS ARE LIKELY TO COME UP?

Sometimes candidates try to guess which questions will be set next year by looking at past papers. They either choose areas of the syllabus that are frequently set or those where questions have not been set in the previous year. For either reason they assume a question could be set the following year. This is a very dangerous thing to do.

- There have been frequent syllabus changes over recent years so no clear pattern can be detected in past papers.
- Examiners set questions on all aspects of the syllabus each year, although the proportions may vary.

Try to master the whole syllabus. When you have tried to answer the questions in these papers, it may help you to identify areas of the syllabus which require revision.

These are areas of the syllabus which Higher tier candidates find difficult.

- **Algebra**, especially solving equations, inequalities and changing the subject of a formula, algebraic fractions, simultaneous equations.
- **Proportion**, particularly inverse proportion.
- **Reversed percentages**, e.g. finding cost before VAT was added.
- **Solving 3-D problems** using Pythagoras and trigonometry.
- **Histograms** and **frequency density**.
- **Conditional probability**, where the probability of an event depends on what has happened before.

## TOP TEN TIPS FOR EXAM SUCCESS

1   Practise all aspects of manipulative algebra, solving equations, rearranging formulas, expanding brackets, factorising, etc.

2   Practise answering questions without the use of a calculator.

3   Practise answering questions with more than one step to the answer, e.g. finding the radius of a sphere with the same volume as a given cone.

4   Make your drawings and graphs neat and accurate.

5   Practise answering questions that ask for an explanation. Your answers should be concise and use mathematical terms where appropriate.

6   Don't forget to check your answers, especially to see that they are reasonable. The mean height of a group of men will not be 187 metres!

7   Lay out your working carefully and concisely. Write down the calculations you are going to make. You usually get marks for showing a correct method.

8   Know what is on, and what is not on, the formula sheet before the examination.

9   Make sure you can use your calculator efficiently. Write down the figures on your calculator and then make a suitable rounding. Don't round the numbers during the calculation. This will often result in an inaccurate answer.

10  Make sure you have read the question carefully so you give the answer the examiner wants!

## HOW TO ACHIEVE A GRADE A IN MATHEMATICS

Contrary to many people's beliefs the award of a grade A is not made each year by awarding it to a fixed percentage of candidates. It is awarded by inspection of the papers and awarding the grade to those candidates whose work has met the required assessment objectives which have been agreed nationally. A brief description of the main assessment objectives is given over the page.

A grade A candidate should be able to:

* give detailed reasons for choices made and justifications and explanations for solutions to problems when investigating within mathematics
* use mathematical language and symbols effectively in presenting a convincing, reasoned argument
* understand and use rational and irrational numbers
* determine the bounds of intervals
* understand and use direct and inverse proportion
* in algebra: rearrange and manipulate formulae; solve equations; find common factors; understand the rules of indices; solve simultaneous equations algebraically and graphically; solve other problems graphically
* use the sine, cosine and tangent of angles of any size
* understand and use Pythagoras' theorem in 2 and 3 dimensions
* understand the conditions for congruent triangles
* calculate the lengths of arcs and the areas of sectors
* calculate the surface area of cylinders and the volumes of cones and spheres
* construct and interpret histograms
* understand different methods of sampling and how they and different sample sizes may affect the reliability of results
* recognise when and how to use conditional probability.

## WHAT EXTRA IS REQUIRED FOR A GRADE A*?

Having established what is required to be awarded a Grade A you might be interested to know what is required for an A* grade. There are at present no A* criteria.

When the Awarding committee is awarding grades they are asked to fix a mark for Grade A and Grade C. Suppose on a particular paper the Grade A mark was fixed at 70 and the grade C mark at 50. These numbers have been chosen only to keep the arithmetic that follows simple. The Grade B mark is then fixed arithmetically half way between 50 and 70, i.e. 60. The Grade A* is then fixed the same number of marks above A as B is below it. In the example we have used A* would be fixed at 80. What does this tell you? Grade A* is a very high standard and relatively few are awarded.

As there are no criteria it is not as clear what examiners are looking for as it is at Grade A.

Generally as the Grade A* boundary is a high mark there is no scope for a bad answer on any paper. A grade A* candidate scores well on all questions.

You should have no weaknesses in any areas of the syllabus.

## HOW TO ASSESS YOUR GRADE

The matrix below suggests grades that you might have expected to achieve with different scores on these papers. These marks are for the two papers in *each* exam and are out of 200. No account has been made of the coursework marks. It is an indication only and does not imply that this is the grade you will receive in the real examination.

Grades D and below are not awarded on Higher Tier.

| A* | 170–200 |
|----|---------|
| A  | 130–169 |
| B  | 90–129  |
| C  | 50–89   |

Letts Educational
The Chiswick Centre
414 Chiswick High Road
London
W4 5TF

Tel: 0845 602 1937
Fax: 020 8742 8767
Email: mail@lettsed.co.uk

Every effort has been made to trace copyright holders and to obtain their permission for the use of copyright material. The authors and publisher will gladly receive any information enabling them to rectify any error or omission in subsequent editions.

First published 2004, 2005
10 9 8 7 6 5 4 3

Text, design and illustrations © Letts Educational 2004

No UK examination boards have supplied or approved the questions, answers or grading advice given in this pack; the answers provided may not be the only solutions to the questions given. The results you achieve in the exams in this pack are only an indication of what you may achieve in the official exam.

Prepared by *specialist* publishing services, Milton Keynes

British Library Cataloguing in Publication Data

A CIP record for this title is available from the British Library

ISBN 1843153084

Letts Educational Limited is a division of Granada Learning Limited, part of Granada plc.

Printed in Great Britain